Counting on Frameworks

Mathematics to Aid the Design of Rigid Structures

Figure 4.30 on page 171 is courtesy of
Kenneth Snelson
American, 1927–
Free Ride Home, 1974
Aluminum and stainless steel
$30' \times 60' \times 60'$
Gift of the Ralph E. Ogden Foundation, Inc.
Storm King Art Center
Mountainville, New York

Library of Congress Catalog Card Number 2001089227
ISBN 0-88385-331-0

Printed in the United States of America

Current Printing (last digit):
10 9 8 7 6 5 4 3 2 1

The Dolciani Mathematical Expositions

NUMBER TWENTY-FIVE

Counting on Frameworks

Mathematics to Aid the Design of Rigid Structures

Jack E. Graver
Syracuse University

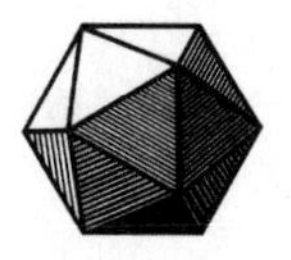

Published and distributed by
The Mathematical Association of America

The DOLCIANI MATHEMATICAL EXPOSITIONS series of the Mathematical Association of America was established through a generous gift to the Association from Mary P. Dolciani, Professor of Mathematics at Hunter College of the City University of New York. In making the gift, Professor Dolciani, herself an exceptionally talented and successful expositor of mathematics, had the purpose of furthering the ideal of excellence in mathematical exposition.

The Association, for its part, was delighted to accept the gracious gesture initiating the revolving fund for this series from one who has served the Association with distinction, both as a member of the Committee on Publications and as a member of the Board of Governors. It was with genuine pleasure that the Board chose to name the series in her honor.

The books in the series are selected for their lucid expository style and stimulating mathematical content. Typically, they contain an ample supply of exercises, many with accompanying solutions. They are intended to be sufficiently elementary for the undergraduate and even the mathematically inclined high-school student to understand and enjoy, but also to be interesting and sometimes challenging to the more advanced mathematician.

1. *Mathematical Gems,* Ross Honsberger
2. *Mathematical Gems II,* Ross Honsberger
3. *Mathematical Morsels,* Ross Honsberger
4. *Mathematical Plums,* Ross Honsberger (ed.)
5. *Great Moments in Mathematics* (*Before 1650*), Howard Eves
6. *Maxima and Minima without Calculus,* Ivan Niven
7. *Great Moments in Mathematics* (*After 1650*), Howard Eves
8. *Map Coloring, Polyhedra, and the Four-Color Problem,* David Barnette
9. *Mathematical Gems III,* Ross Honsberger
10. *More Mathematical Morsels,* Ross Honsberger
11. *Old and New Unsolved Problems in Plane Geometry and Number Theory,* Victor Klee and Stan Wagon
12. *Problems for Mathematicians, Young and Old,* Paul R. Halmos
13. *Excursions in Calculus: An Interplay of the Continuous and the Discrete,* Robert M. Young
14. *The Wohascum County Problem Book,* George T. Gilbert, Mark Krusemeyer, and Loren C. Larson
15. *Lion Hunting and Other Mathematical Pursuits: A Collection of Mathematics, Verse, and Stories by Ralph P. Boas, Jr.,* edited by Gerald L. Alexanderson and Dale H. Mugler
16. *Linear Algebra Problem Book,* Paul R. Halmos

17. *From Erdős to Kiev: Problems of Olympiad Caliber,* Ross Honsberger
18. *Which Way Did the Bicycle Go? ... and Other Intriguing Mathematical Mysteries,* Joseph D. E. Konhauser, Dan Velleman, and Stan Wagon
19. *In Pólya's Footsteps: Miscellaneous Problems and Essays,* Ross Honsberger
20. *Diophantus and Diophantine Equations,* I. G. Bashmakova (Updated by Joseph Silverman and translated by Abe Shenitzer)
21. *Logic as Algebra,* Paul Halmos and Steven Givant
22. *Euler: The Master of Us All,* William Dunham
23. *The Beginnings and Evolution of Algebra,* I. G. Bashmakova and G. S. Smirnova (Translated by Abe Shenitzer)
24. *Mathematical Chestnuts from Around the World,* Ross Honsberger
25. *Counting on Frameworks: Mathematics to Aid the Design of Rigid Structures,* Jack E. Graver

MAA Service Center
P.O. Box 91112
Washington, DC 20090-1112
1-800-331-1MAA FAX: 1-301-206-9789

Preface

The formal name for the topic discussed in this book is *rigidity theory*. It is a body of mathematics developed to aid in designing structures. To illustrate, consider a scaffolding that has been constructed by bolting together rods and beams. Before using it, we must ask the crucial question: *Is it sturdy enough to hold the workers and their equipment?*

Several features of the structure have to be considered in answering this question. The rods and beams must be strong enough to hold the weight and withstand a variety of stresses. In addition, the way they are put together is important. The bolts at the joints must hold the structure together, but they are not expected to prevent the rods and beams from rotating about the joints. So the sturdiness of the structure depends on the way it has been braced. Just how to design a properly braced scaffolding (or the structural skeleton of any construction) is the problem that motivates rigidity theory.

Thinking further about the scaffolding problem, we might conclude that it really has two separate parts. Designing the optimal placement of the rods, beams and braces might require mathematics that is quite different from what is needed to analyze the strength and weight-bearing properties of the materials used. In fact, we would like to separate these two aspects of structure building.

Our ultimate aim is to be able to take a very rough sketch of a proposed scaffolding—a sketch without reference to scale, types of materials or exact alignments of its members—and decide whether the final structure will be rigid, assuming that the other aspects (strength of materials etc.) are properly attended to.

The information depicted in this rough sketch is called the *combinatorial* information about the structure: the way various elements are to be *combined* to form the structure. The combinatorial information is in contrast to the *geometrical* information, the exact lengths and locations of the various elements, and to the remaining information about the materials to be used.

The purpose of this book is to develop a mathematical model for rigidity. We will actually develop three distinct models. The structure we wish to study will be represented by a *framework*: a configuration of straight line segments joined together at their endpoints. We will assume that the joints are flexible and will explore whether a given configuration can be deformed without changing the lengths of its segments. For instance, a triangle is a simple rigid framework in the plane, while the square is not rigid—it deforms into a rhombus.

The first model we will develop for rigidity is the *degrees of freedom* model. It is very intuitive and very easy to use with small frameworks. But at the outset this model lacks a rigorous foundation, and in some cases it actually fails to give a correct prediction.

The next model we will build is based on systems of quadratic equations. Each segment of the framework yields the quadratic equation stating that the distance between the joints at the segment's two ends must equal the length of the segment itself. It is rather easy to see that this model will give the correct answer to our question about the presence or absence of rigidity, in all cases. Hence, the quadratic model is the standard model of rigidity. Unfortunately, this model is not so easy to use even with small frameworks.

The third model we will construct is equally accurate but is based on a slightly different definition called *infinitesimal rigidity*. In this model, the quadratic equations are replaced with linear equations, and it is therefore much easier to use. The two models do not give identical predictions. But, as we shall see, if a framework is infinitesimally rigid (based on the linear model), it is rigid (based on the quadratic model). A few exceptional frameworks exist that are rigid but not infinitesimally rigid; however, this fact is a small price to pay for the ease of using the linear model.

The quadratic and linear models for rigidity and infinitesimal rigidity are both combinatorial and geometric in nature; they depend on the exact lengths and locations of the segments of the framework as well as the way the segments are put together. The degrees of freedom model, on the other hand, is purely combinatorial. So, after developing infinitesimal rigidity, we will return to this simple degrees of freedom model. Using infinitesimal rigidity, we will be able to justify the degrees of freedom model for a very large class of frameworks. In fact, we will show that all three models agree except for very few very special frameworks. We will end our discussion by developing the degrees of freedom model in purely combinatorial terms for the plane and considering the extension of this development to 3-space.

Chapter 1 introduces rigidity through a rather special problem. Our first model, degrees of freedom, is introduced here. We develop this model as far as we can without introducing specialized combinatorial tools.

In Chapter 2, we investigate a very small portion of a very large field called *graph theory*. Along the way we introduce our second (quadratic) model for rigidity.

The level of rigor is ratcheted up a bit in Chapter 2 and then again in Chapter 3. There rigidity (the quadratic model) is developed further and infinitesimal rigidity (the linear model) is introduced. Returning to the degrees of freedom model, we use what we have learned about infinitesimal rigidity to show that the degrees of freedom model is purely combinatorial and that it is applicable to "almost all" frameworks.

Showing that the degrees of freedom model is combinatorial is one thing; understanding how to use this fact is another. Even if we know that the rough diagram of our scaffolding contains all of the information needed to decide whether the scaffolding is rigid or not, that is of little use if we don't know how to make the decision! We will develop several techniques for making this decision for planar frameworks and, in the last section of Chapter 3, discuss their extensions to 3-space.

The first section of Chapter 4 is devoted to a short outline of the history of rigidity, which started with a conjecture of Euler. In the second section, we consider frameworks from a different point of view. Instead of designing frameworks that are rigid, the problem is to design frameworks that deform in some special way. Frameworks that deform are often called *mechanisms*. The designing of mechanisms flourished during the industrial revolution. We will highlight the long search for a mechanism that yields linear motion. Such a mechanism can replace a cam sliding in a slot and thereby reduce friction. Designing geodesic domes is the next topic of Chapter 4, and the chapter ends with an investigation of *tensegrity*, rod and cable frameworks that achieve rigidity by being put under tension.

Relatively little mathematical background is needed to read Chapter 1 and, for that matter, much of Chapter 4. For Chapter 2, comfort with abstract mathematical concepts and with simple proofs is needed. In particular, there are several inductive proofs. We also use some concepts from elementary calculus in defining our second model of rigidity. In Chapter 3, we introduce infinitesimal rigidity, which requires a basic understanding of the vector space R^n. Chapter 3 also includes several proofs involving calculus and linear algebra. However, it is possible for the reader to skip those more involved proofs but still have the necessary background for a complete reading of Chapter 4.

Finally, an annotated bibliography gives information as to where the reader may find more extensive developments of the topics touched upon in the book.

Within each chapter, figures are numbered consecutively; so are the exercises. Likewise the group of items comprising lemmas, theorems, corollaries, conjectures and results is sequentially numbered. Of the 90 or so exercises sprinkled throughout the book, only a few can be called routine. The exercises are intended to illuminate the topic under discussion or to lead the reader into one of the many interesting side trips that present themselves. The exercises vary in difficulty; the more difficult ones usually include some indication as to their level of difficulty and frequently include a hint.

I am greatly indebted to the referees from the editorial board of the Dolciani Series for their many valuable suggestions and their probing questions. I owe a special thanks to Dan Velleman for many thoughtful suggestions. I also thank Greg Vassallo, who took all of the photographs except the photograph of the Kenneth Snelsen sculpture. That was supplied by the Storm King Art Center. Finally, many thanks to my wife, Yana, without whose support the book would have never been written.

—J.E.G.

Contents

CHAPTER 1

Counting on Frameworks

1.1 Designing Structures

Think about building something. The process of building frequently starts like this: First, we have a mental image of the object we wish to build; next, we make a rough sketch; finally, after several more intervening steps, the object, or perhaps a prototype, is constructed. At this stage we can test whether or not the object has all of the properties that our mental image possessed. Discovering at this stage that one or more critical properties is not satisfied is somewhat of a disaster, involving a great deal of wasted effort. Careful planning is essential to reducing the probability of such disasters.

We will restrict our attention to building structures of a special type and will concentrate on the planning stage. To be specific, the structures we will consider are constructed out of rigid rods connected together at their ends. Look at the cube sketched in Figure 1.1. The "rigidity" of this structure depends entirely on the strength of the joints in maintaining the angles between the rods. Under enough pressure the faces will distort and the structure will collapse. The rigidity of the tetrahedron, on the other hand, does not depend on the joints in this way. Its joints can be flexible and the tetrahedron will still be rigid.

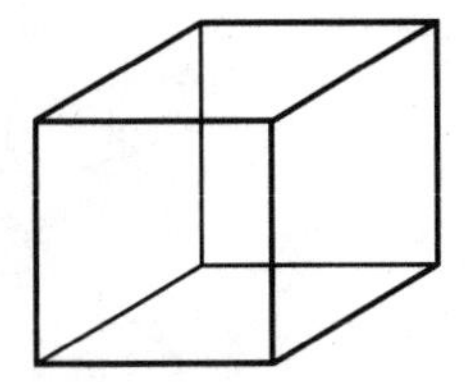

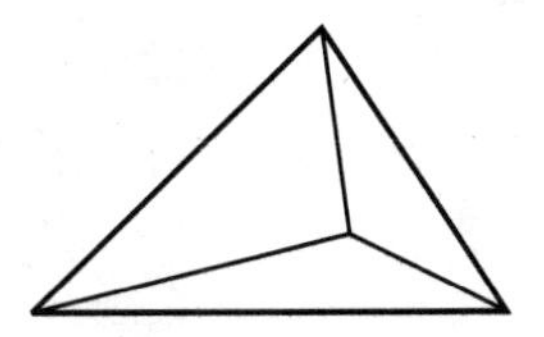

FIGURE 1.1

It is in this sense that we will use the term rigid: A structure made from rods will be rigid if it resists distortion even when all of its joints are flexible. And herein lies our problem:

> *Can we accurately predict from the rough sketch alone whether or not the finished structure will be rigid?*

With the cube and tetrahedron, we are confident in our predictions. But, having accurately predicted that the unbraced cube is not rigid, suppose that we would like to add sufficient braces to make it rigid. How should we proceed?

In Figure 1.2, we have drawn three different bracing schemes. We can argue that the left-hand bracing results in a rigid cube: It is made up of five tetrahedra, one in the center whose edges are the six braces and four other tetrahedra, each sharing a face with the central one. In the second proposed bracing, the braces in the top and bottom faces of the cube have been replaced by the opposite diagonals. What can we say now? With no tetrahedra in sight, can we be sure of any prediction we may make? The third proposed bracing is even more problematic. We have to be careful here to be sure that we understand what the picture is intended to portray. Since we are bracing the cube, no further joints have been introduced and the diagonal braces cross one another but are not connected; they would be permitted to slide across each other if this braced cube were not rigid.

Exercise 1.1. *Before you read any further, make your prediction for the rigidity of the second and third structures sketched in the figure.*

To make a "practical" problem out of these examples think of such a cube as a base for a glass topped coffee table. The rods and rubber joints can be easily mailed to the consumer who can then put them together, purchasing the glass top locally. In the spirit of this interpretation, let's answer some of our questions by building a model.

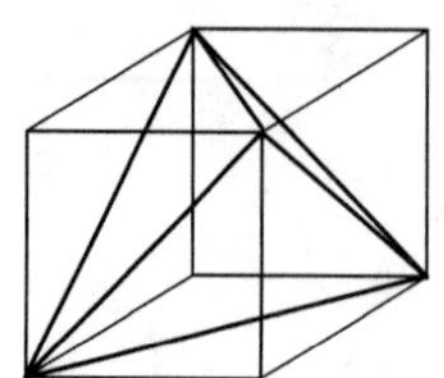
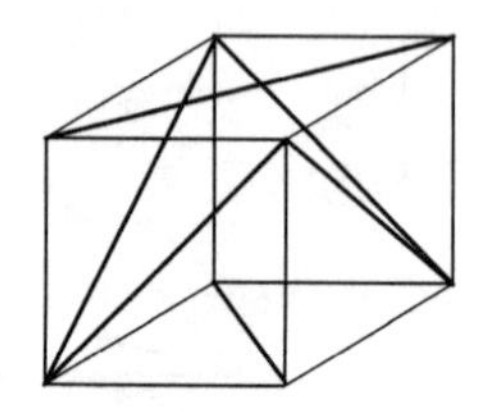
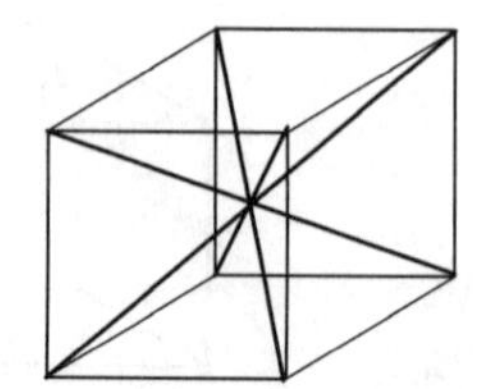

FIGURE 1.2

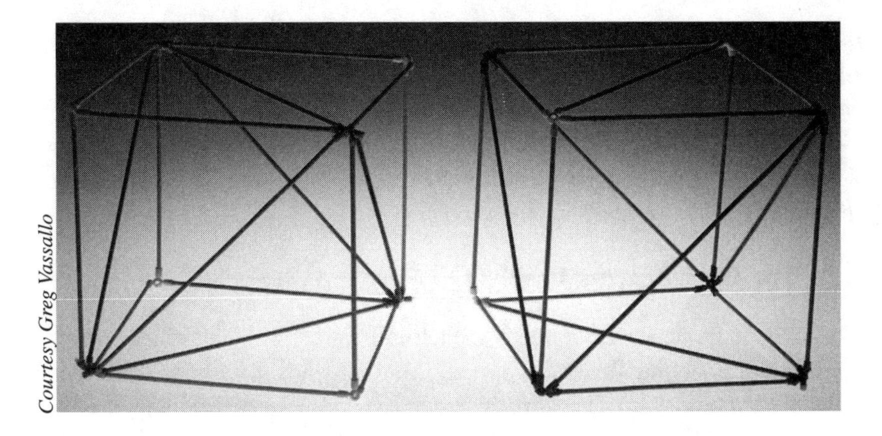

Courtesy Greg Vassallo

FIGURE 1.3

Models of the first two proposed bracings are easy to construct. Using a standard model building kit, we have built these models and photographed them (Figure 1.3). If you were to build such models yourself, you would verify directly that they are rigid.

It is not so easy to build a model based on the third sketch. Actual rods have thickness, and at the crossing point in the center they must be bent slightly. So it is impossible to build an undistorted model. On the other hand, if you did manage to build a model that was not too badly distorted, it would seem to be rigid. However, it would not seem to have the robust rigidity of the first two braced cubes, and you might conclude that a coffee table of this design is not such a good idea. You might also conclude that there should be a better way to decide rigidity questions than by building models. In fact, rigidity is clearly one of those properties we would like to be sure of long before we get to the building stage.

The main purpose of this book is to explore the mathematical theory of rigidity and to develop methods for predicting rigidity early in the planning stage of a structure. The only mathematical knowledge expected of the reader is that usually covered in a standard calculus sequence and a first course in linear algebra (an introduction to the vector space of n-tuples of real numbers, to be specific). Because of this limitation on the required background, the reader will be asked to accept a few of the results without proof. But even these missing proofs can be motivated and explained in part with the tools available to us.

While the rigidity of 3-dimensional structures is the initial stimulus for this study, we will spend a lot of time and effort considering rigidity in lower dimensions. Just as an extensive knowledge of plane geometry is necessary for a good understanding of solid geometry, an extensive knowledge of plane rigidity is necessary for a good understanding of spatial rigidity. In fact, 1-dimensional rigidity will play an important role in our development. In the next section, we start our investigation by considering a very special problem in planar rigidity: the grid bracing problem.

1.2 The Grid Bracing Problem

Consider the framework pictured at the left in Figure 1.4. Think of it as being constructed by bolting four rods together near their ends. Even if we carefully tighten the bolts, the framework will deform under sufficient pressure, as pictured next. If we wish to prevent this deformation and make the framework rigid, we may include a diagonal brace, as illustrated in the third part of the figure. (Of course, it makes no difference which of the two diagonals we include.)

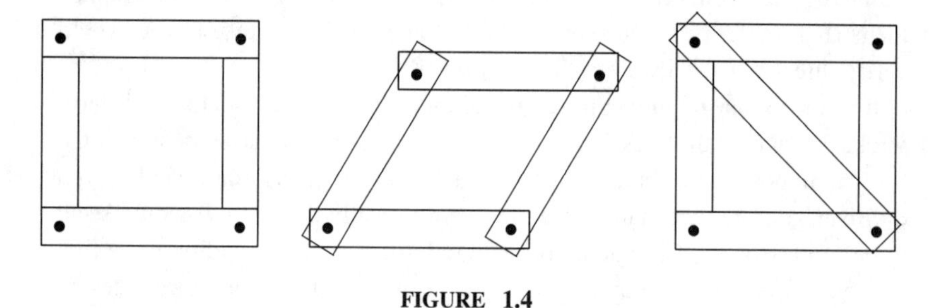

FIGURE 1.4

Now consider a rectangular grid consisting of several such squares. In Figure 1.5, we diagram such a framework with a simpler sketch, where the joints are represented by dots and the rods by line segments. We will refer to such a framework by the number of its vertical and horizontal squares or cells. Thus, the framework diagrammed in Figure 1.5 is the 3×5-*grid*. Such

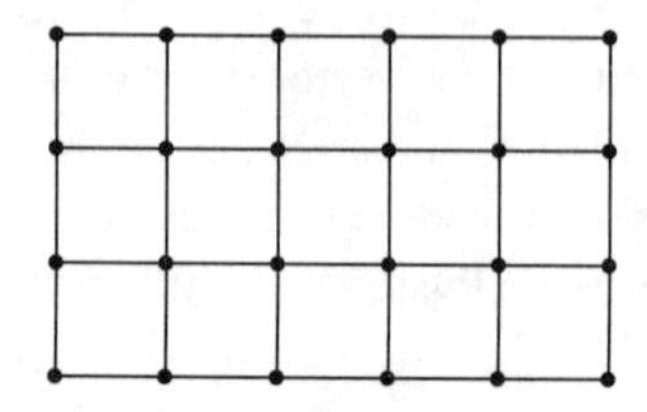

FIGURE 1.5

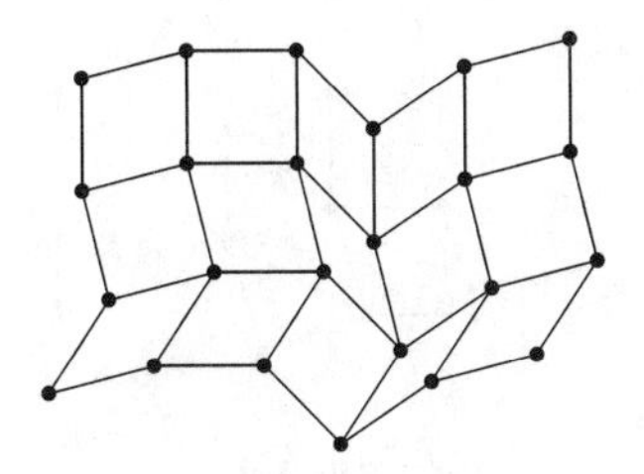

FIGURE 1.6

a grid may be deformed in many ways. One such deformation is illustrated in Figure 1.6.

The basic problem is to brace enough of the cells so that the braced grid will be rigid. We could brace every cell. But, as we will see, for all but the simplest of grids, this is unnecessary—a much smaller number of braces will do. Hence, the natural questions to ask are: "How many braces are needed to make a given grid rigid?" and "In which cells should those braces be placed?" We can start our investigation with a very simple, but nevertheless useful, observation:

If all of the cells in any column or row of a grid are unbraced, the grid is not rigid.

This observation is illustrated in Figure 1.7.

It follows from this observation that all cells in the simplest grids must include braces:

To be rigid, the $1 \times k$*-grid or the* $k \times 1$*-grid must have all cells braced!*

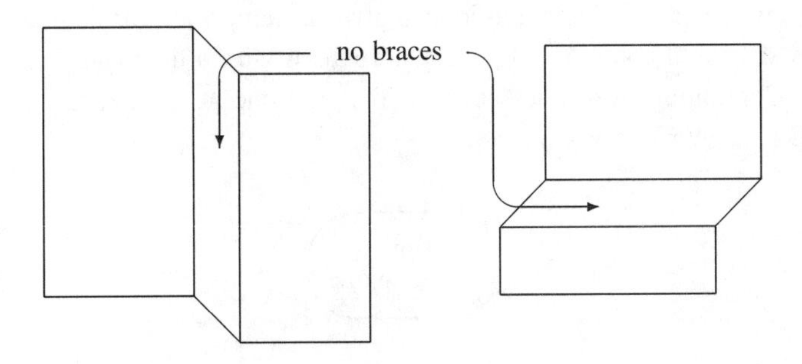

FIGURE 1.7

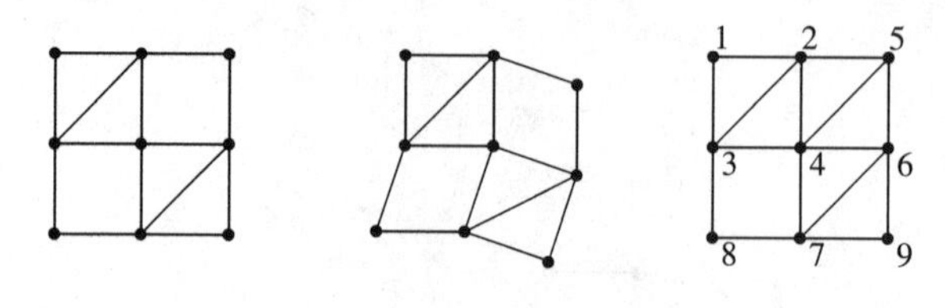

FIGURE 1.8

Thus, we have solved the grid bracing problem for $1 \times k$-grids and the $k \times 1$-grids. We may also conclude that a $2 \times k$-grid (a $k \times 2$-grid) will need at least k braces to be made rigid.

Consider the 2×2-grid. The only arrangements of two braces that leave no row or column unbraced are those that brace diagonally opposite cells. The first picture of Figure 1.8 exhibits such a bracing and the second picture illustrates that such a bracing does not induce rigidity. The third part of that figure depicts the only way, up to symmetry, to select three cells to brace. We would like to demonstrate that this last bracing is rigid. We can do this using a basic observation about planar rigidity in general.

Suppose that we have an arbitrary rigid planar framework consisting of rods bolted together at their ends (not necessarily a braced grid). It should be clear that, if we attach the free ends of two new joined rods, the resulting framework will also be rigid. See Figure 1.9.

In other words:

> *Attaching two rods with a common joint to a rigid framework results in another rigid framework.*

Using this observation, we may verify that the right-hand braced grid in Figure 1.8 is rigid. Start with the single edge (with endpoints labeled 1 and 2 in Figure 1.8) and attach joint 3 by two rods to get a triangle. Clearly, a single triangle is rigid. Attaching joint 4 gives a framework consisting of two triangles sharing a common edge (a quadrilateral with a diagonal) and it too is rigid. Continuing, we attach joints $5, 6, \ldots, 9$ one at a time, to construct the third framework in Figure 1.8.

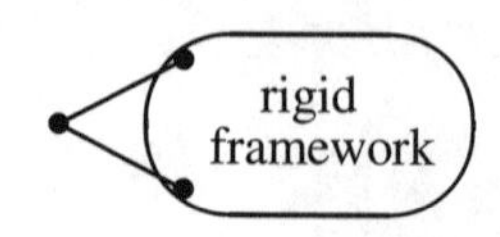

FIGURE 1.9

One may build a wide variety of rigid frameworks using this technique. The simplest is the triangle; next, with four joints, is the quadrilateral with diagonal. We say "the triangle" although of course there are infinitely many different triangles. We are interested only in the properties that all triangles have in common, among them "being rigid." It is in this generic sense that we use the term "the triangle." In the same way, we speak of "the quadrilateral with diagonal" no matter how it may be pictured; see Figure 1.10.

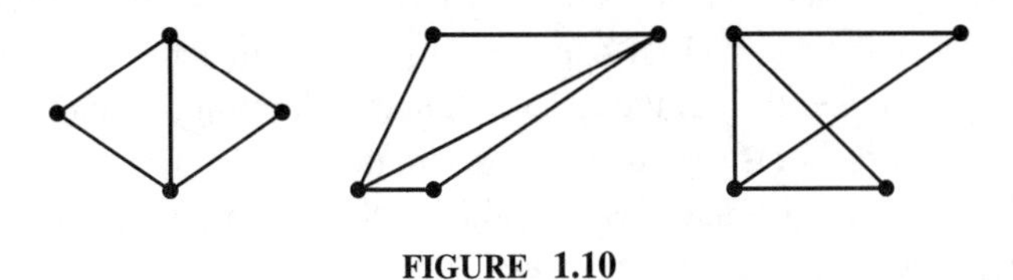

FIGURE 1.10

Continuing with our building project, there are three fundamentally different ways to attach a new joint by two edges to the quadrilateral with diagonal. They are pictured in Figure 1.11 with the new joint circled. The number of different rigid frameworks produced by this method grows very rapidly as n, the number of joints, gets large and many rigidly braced grids are included among them.

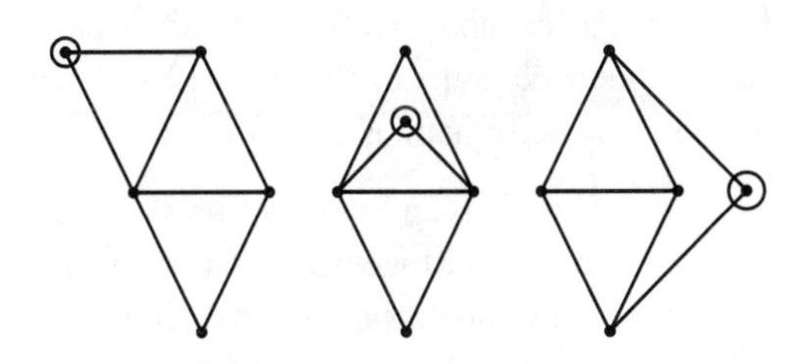

FIGURE 1.11

Exercise 1.1. *Construct all 11 fundamentally different rigid frameworks with six joints that may be produced by this method.*

This exercise should prove very challenging and enlightening for the reader who attempts it. In tackling this problem, the reader will face two issues common to most combinatorial searches:

- how to be sure to get them all;
- how to recognize duplicates that are just drawn differently.

To check whether a planar framework is the result of such a construction, we reverse the construction process: Look for a joint connecting just two rods; remove it and the rods; then repeat this step as often as is possible. If the final step results in a single edge, then the framework can be rebuilt by this construction. We can actually apply this reduction process to any framework that has a joint connecting just two rods. So let's take a closer look.

To simplify the discussion, we make a definition: A joint connecting k rods is said to have *valence* k. If a framework has a joint of valence 2, we may remove it (the joint and rods it connects) to get a simpler framework. We have already observed that, if the simpler framework is rigid, then the larger, original framework is also rigid. But what if the simpler framework is not rigid? What can we conclude then? Could attaching two joined rods make a nonrigid framework into a rigid framework?

To answer this last question, we introduce some notation and then consider two cases. First the notation: Denote the joint connecting the two rods we are adding by $\mathbf{p}$ and the joints at the other ends of the two rods by $\mathbf{q}$ and $\mathbf{r}$. Also denote the nonrigid framework to which we are attaching rods by $\mathcal{F}$.

- Case 1. Now, $\mathcal{F}$ will have parts that are rigid. Certainly the rods of $\mathcal{F}$ are rigid, and there may be even bigger parts that are rigid. If $\mathbf{q}$ and $\mathbf{r}$ lie on the same one of these rigid parts, then $\mathbf{p}$ and its rods are simply attached to that part and cannot interfere with any of the deformations of $\mathcal{F}$. (See the left-hand diagram in Figure 1.12.) Hence the enlarged framework has the same deformations and is not rigid.

- Case 2. If, on the other hand, $\mathbf{q}$ and $\mathbf{r}$ do not lie on the same rigid piece of $\mathcal{F}$, then they may move closer together or farther apart as $\mathcal{F}$ deforms. If $\mathbf{p}$ is attached so that the angle between the rods at $\mathbf{p}$ is neither 0 degrees nor 180 degrees, then, at least small deformations of $\mathcal{F}$ will not be hampered; the angle at $\mathbf{p}$ will simply change as $\mathbf{q}$ and $\mathbf{r}$ move closer or farther apart. (See the right-hand diagram in Figure 1.12.) Thus, in this case too, the enlarged framework has deformations and is not rigid.

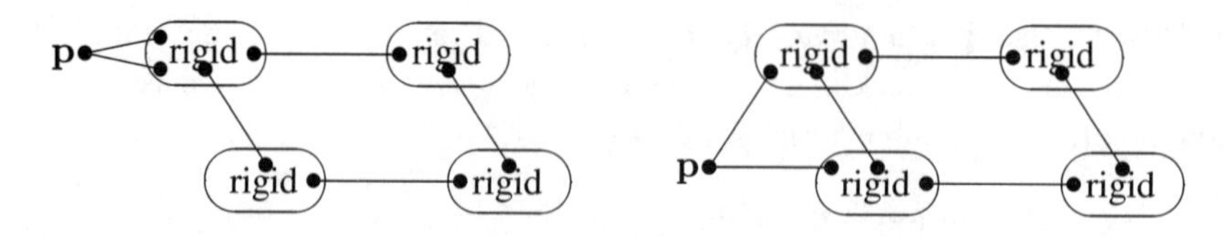

FIGURE 1.12

We may conclude:

1. *If a planar framework $\mathcal{F}'$ is constructed from a rigid framework $\mathcal{F}$ by attaching a new joint by two rods, then $\mathcal{F}'$ is rigid.*
2. *If a planar framework $\mathcal{F}'$ is constructed from a nonrigid framework $\mathcal{F}$ by attaching a new joint by two rods such that the angle between the attaching rods is neither* 0 *degrees nor* 180 *degrees, then $\mathcal{F}'$ is nonrigid.*

Thus, a joint of valence 2 such that the angle between the attaching rods is neither 0 degrees nor 180 degrees is called a *removable joint.* And we note that any framework with one or more removable joints may be reduced to a simpler framework by stripping off the removable joints. If we can then decide whether the simpler framework is rigid or not, we can decide whether the original framework is rigid or not.

Another important observation can be made about this construction:

> *A rigid framework constructed from a single edge by repeatedly attaching pairs of joined rods satisfies the equation $r = 2n - 3$, where r denotes the number of rods and n the number of joints.*

This equation holds for a single rod ($n = 2$ and $r = 1 = 2 \times 2 - 3$) and the triangle ($n = 3$ and $r = 3 = 2 \times 3 - 3$). Since at each step in the construction r increases by 2 and $(2n - 3)$ increases by 2, the formula holds in all cases.

Let's see what this observation tells us about any braced $h \times k$-grid that can be constructed in this way. Let n denote the number of connecting joints in the framework. It is easy to see that $n = (h+1)(k+1)$. To compute r', the number of rods in the grid itself, it is best to count the horizontal and vertical rods separately. There are $(h+1)k$ horizontal rods and $h(k+1)$ vertical rods, giving a total of $r' = 2hk + h + k$. If b is the number of braces, we have $r = r' + b$ and

$$\begin{aligned} b &= r - r' \\ &= (2n - 3) - (2hk + h + k) \\ &= 2(h + 1)(k + 1) - 3 - (2hk + h + k) \\ &= h + k - 1. \end{aligned}$$

This agrees with what we already know about the $1 \times k$-grids: They need $1 + k - 1 = k$ braces. For instance, the rigid 2×2-grid in Figure 1.8 has $2 + 2 - 1 = 3$ braces.

We can conclude:

Any rigid $2 \times k$-grid constructed by this method would have to have $k + 1$ braces.

Remembering that each column must have at least one cell braced, we conclude further:

Exactly one column has both cells braced, and that every other column has exactly one cell braced.

This is, in fact, an accurate description of all rigid $2 \times k$-grids using a minimum number of braces. Using only the observations developed so far, the interested reader can now solve the $2 \times k$-grid bracing problem.

Exercise 1.2. *Prove that a $2 \times k$-grid with both cells braced in one column and one cell braced in every other column is rigid and, furthermore, that any $2 \times k$-grid with fewer than $k + 1$ cells braced is not rigid.*

Even though the arguments used above apply only to braced grids that can be constructed in this very special way, it is tempting to make the following conjecture.

Conjecture 1.1. *The $h \times k$-grid needs to have at least $h + k - 1$ of its cells braced and, if properly placed, $h + k - 1$ braces will suffice to make it rigid.*

With this conjecture in mind, consider the bracings of the 3×3-grid in Figure 1.13. Since $3 + 3 - 1 = 5$, we would expect that the first framework has enough braces to be rigid. Indeed it is rigid: removing the four corner joints uncovers four more 2-valent joints, and their removal results in another four 2-valent joints; removing these last four 2-valent joints leaves a square with a diagonal which is, of course, rigid. Hence, rebuilding this first braced grid, we see that at each step it remains rigid.

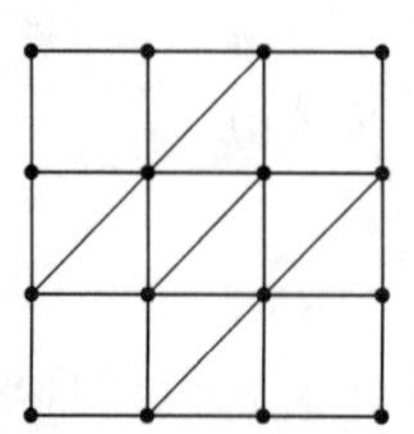
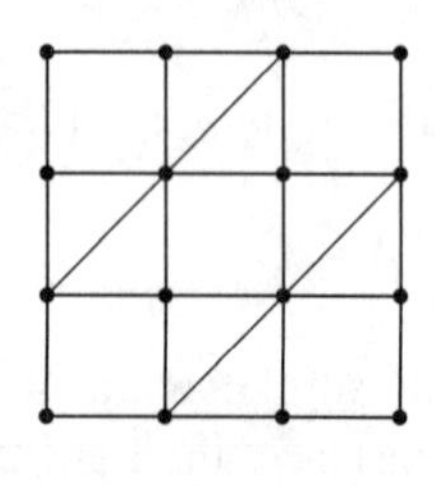
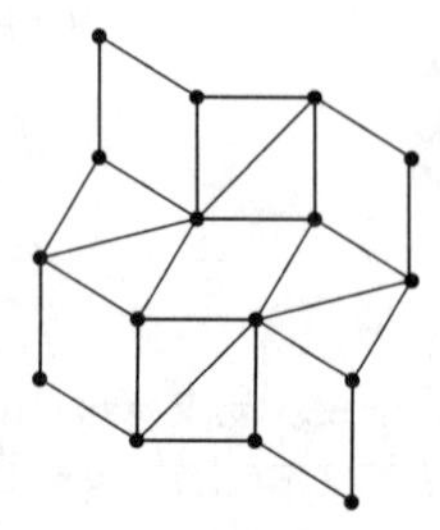

FIGURE 1.13

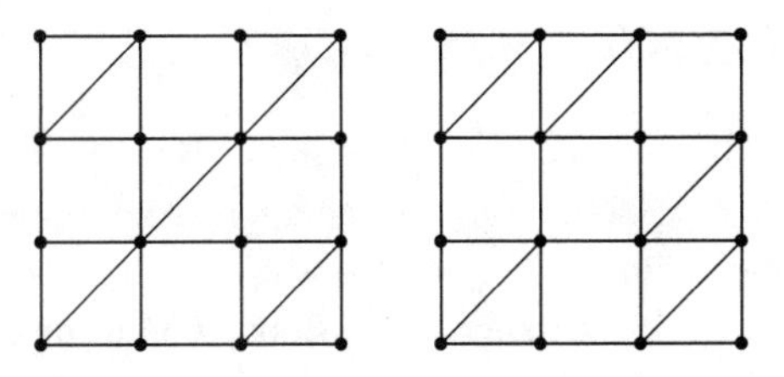

FIGURE 1.14

The same deconstruction applied to the second example works down to the square without a diagonal. We conclude that the second bracing of the grid (which has only four braces) is not rigid. To see a distortion of this second braced grid, distort the inner square before rebuilding the grid. We illustrate the resulting deformation to the right.

Now consider the two braced grids in Figure 1.14. Both have the predicted five braces. But, with the techniques developed so far, we cannot decide whether they are rigid or not. After removing the two 2-valent joints from the first grid or the three 2-valent joints from the second grid, the deconstruction stops: All remaining joints have valence 3 or more. These reduced frameworks are almost as complicated as the original ones.

The fact is that one of these braced grids is rigid and the other is not. But unless your geometric skills are in excellent condition, you probably will not be able to visualize the distortion that will show one to be nonrigid nor construct an argument to verify that the other is rigid. (It may be very instructive, nonetheless, to try.)

So it is clear that we are going to need some more powerful tools to investigate larger grids. These tools are more naturally developed in the "framework" of frameworks in general. Our task in the next section is to set the stage for this development by taking a closer look at the general theory of frameworks.

Exercise 1.3. *Consider the $h \times k$-grid. Select one row and one column, then brace every cell in that row and column. Now verify each of the following statements.*

1. *There are $h + k - 1$ braces.*
2. *This bracing makes the grid rigid.*
3. *If any brace is deleted, the grid is no longer rigid.*

Exercise 1.4. *Analyze the following bracing scheme for the $h \times k$-grid: Brace all cells in a row (column) and exactly one cell in every other row (column).*

1.3 Frameworks and Their Motions

The concept of a framework is very intuitive. But any detailed study will require a foundation of definitions and notation. And while we are building this "mathematical framework" for the study of planar frameworks, we will include frameworks in other dimensions. For instance, 3-dimensional frameworks, geodesic domes and the like, are of great interest. Also, 1-dimensional frameworks are more interesting than one might suppose.

With these generalizations in mind, let us proceed to refine our concept of a framework. The final, formal definition will not come until Chapter 2, however. We begin by specifying:

- A *planar (linear, spatial) framework* consists of a finite set of points in the plane (line, 3-space) some pairs of which are joined by line segments.

Henceforth we will use these more common mathematical terms *point* and *segment*, leaving *joint* and *rod* for when we speak of a physical manifestation of a framework.

In our quest for rigidity, the next concept to consider is what is meant by a *motion* of a framework. Intuitively, we may think about moving the points of the framework so that the lengths of segments of the framework remain unchanged throughout the motion. Furthermore, the entire motion should take place in the embedding space; in other words, the motions of a planar (linear, spatial) framework must take place in the plane (line, 3-space).

In Figure 1.15, we illustrate two kinds of motions of a planar framework.

First, consider the left-hand framework and think of rotating the segment joining $\mathbf{p}$ and $\mathbf{q}$ clockwise about the point $\mathbf{q}$. The results of this motion are pictured in the next two frameworks in the figure. Notice that while preserving the lengths of all of the segments of the framework, these motions do alter distances between some points of the framework—for example, the distance between $\mathbf{p}$ and $\mathbf{r}$ changes from sketch to sketch.

This observation helps us distinguish between two types of motions:

- A motion of a framework, in any dimension, is called a *deformation* of the framework if it preserves the lengths of all of the segments of the framework but alters the distance between some pair of points of the framework.

- Motions that preserve distances between all pairs of points of the framework are called *rigid motions* of the framework.

Rotating the entire framework 180 degrees about the point c is a rigid

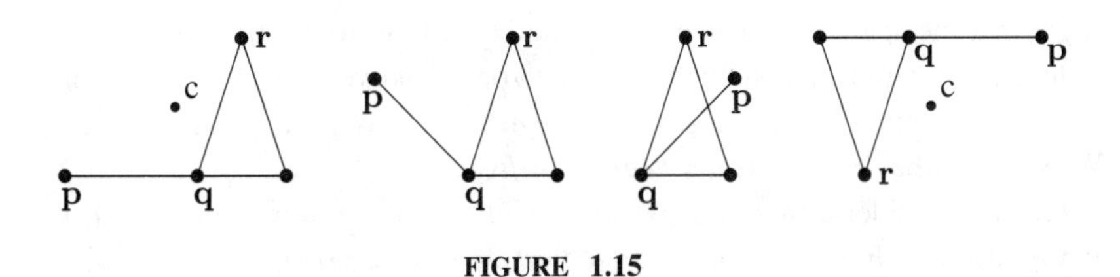

FIGURE 1.15

motion. The result of this *half-turn* is illustrated by the framework on the far right in Figure 1.15.

The rigid motions of a planar framework may be described as the rigid motions of the plane restricted to the framework. One way to visualize a rigid motion of a framework is to picture copying the framework on tracing paper and then sliding the tracing paper into some other position. The sliding of the tracing paper represents a rigid motion of the entire plane. It is important to note that the entire motion must take place in the plane. This limitation rules out other mappings of the plane onto itself, such as reflections, even though they preserve all distances.

Finally:

- We define a framework to be *rigid* if it admits no deformations, that is, if all of its motions are rigid motions.

We should note that whether or not a framework is rigid depends very much on the dimension of the embedding space. As a planar framework, the rectangle with diagonal is rigid; but, as a spatial framework, it admits a deformation: It folds like a hinge along the diagonal brace. Similarly, a framework consisting of two segments with a common endpoint is not rigid in the plane; but, when restricted to a line, its only motions are translations and it is rigid.

1.4 Degrees of Freedom

Consider a point in the plane. It can be moved horizontally and vertically. Furthermore, the new position resulting from any motion of the point may be achieved by a combination of a horizontal and a vertical motion. For this reason, we say that the point has 2 *degrees of freedom.* Another way to think of this is to coordinatize the plane and to note that it then takes two real numbers to identify the location of the point.

Next consider a line segment of length l in the plane; let $\mathbf{p}$ and $\mathbf{q}$ denote its endpoints. We need three pieces of information to identify its position in the plane: One possibility is the two coordinates of $\mathbf{p}$ and the counterclockwise angle that the segment makes with the horizontal ray to the right through $\mathbf{p}$. In terms of motions, a combination of horizontal and vertical translations and a rotation about its center can place a segment in any position in the plane. We say that the segment has 3 *degrees of freedom.*

Finally, consider a more complicated rigid object, say a triangle or a disk. If we choose a line segment on the object and if we do not turn the object over (reflect it), its position in the plane is given by the three numbers that locate the segment. So we say that any rigid body in the plane, except a single point, has 3 *degrees of freedom in the plane.*

We can use this simple idea of degrees of freedom to analyze frameworks in the plane. Let's start with a framework consisting of only two points and no segments:

1. Each point can move independently, so the framework has 4 degrees of freedom.
2. As we have seen, adding a segment between the two points would reduce the number of degrees of freedom to 3.

In this case, therefore, adding a segment reduces the number of degrees of freedom by 1. Next, let's consider the triangle:

1. We start with 6 degrees of freedom for the three points without any segments. This is pictured on the left in Figure 1.16.
2. Adding a segment reduces the total number of degrees of freedom to 5 to 2 for the isolated point and 3 for the segment. (The rotational degree of freedom of the segment is denoted by the circle in the figure.)
3. Adding a second segment yields 4 degrees of freedom: 3 for the first segment plus 1 for the rotation of the second segment about their common endpoint.
4. Finally, the third segment reduces the number of degrees of freedom to 3, which is correct for the rigid triangle.

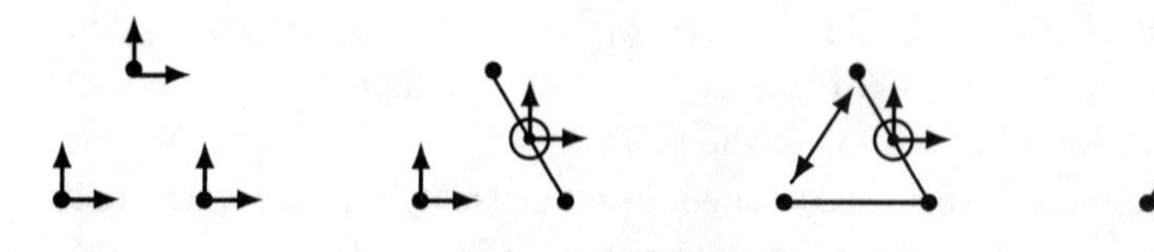

FIGURE 1.16

For our purposes of designing and evaluating constructions, it is important to ask how rigorous such degrees of freedom arguments can be. In fact, we want to ask even more specifically:

> *How rigorous can we make the definition of the degrees of freedom of a complex framework in the plane?*

Let's take the following as a working definition:

- A planar framework $\mathcal{F}$ with more than one point has k internal degrees of freedom and $k+3$ (total) degrees of freedom, where k is the number of segments that must be added to $\mathcal{F}$ to make it rigid.

To illustrate the use of this definition let's take another look at our grids. Consider the quadrilateral (the 1×1 grid):

1. Here we have four points; so starting with no segments, we have 8 degrees of freedom (5 internal degrees of freedom).
2. The four segments then decrease this to 4 degrees of freedom (1 internal degree of freedom).
3. The quadrilateral has the 3 degrees of freedom of any rigid framework (2 translational and 1 rotational); the 1 internal degree of freedom corresponds to the deformation of the framework, as illustrated in Figure 1.4.
4. If we now add one more segment (a diagonal), the number of degrees of freedom of the framework is reduced to 3, signifying that it is now rigid.

We should note at this point that adding a sixth segment cannot reduce the number of degrees of freedom any further. Adding segments to a rigid part of a framework has no effect on the number of degrees of freedom of the framework itself. In general:

> *If there is no deformation of a framework that alters the distance between two of its points, adding a segment between those points has no effect on the number of degrees of freedom of that framework.*

Such a segment is called an *implied segment.*

It seems, then, that the number of internal degrees of freedom of a framework can be computed by adding segments one at a time until the resulting framework is rigid, making sure at each step that the segment to be added is not an implied segment. The number of segments added is then the number of internal degrees of freedom of the original framework. Thus, we expect that a framework with n points and r *properly placed* segments will be rigid whenever the number of the segments is sufficient to bring down the total number of degrees of freedom from the initial count of $2n$ (points only) to 3 (the number of degrees of freedom of a rigid framework). Setting $2n - r$ equal to 3 and solving for r, we conclude:

A framework with n joints should be made rigid with $2n - 3$ appropriately placed segments.

We have already justified this formula for those frameworks that are constructible by adding a sequence of 2-valent points. In fact, this intuitive degrees of freedom approach works for most planar frameworks. Its success depends on the assumption that adding an edge can reduce the number of degrees of freedom count by at most one. However, in some very special cases, adding an edge may decrease the degrees of freedom count by more than one.

Consider the left-hand framework pictured in Figure 1.17. With $n = 6$ and $r = 7$, we would assume that it had $12 - 7 = 5$ degrees of freedom. Adding two segments, one joining $\mathbf{q}$ to $\mathbf{t}$ and the other joining $\mathbf{t}$ to $\mathbf{r}$, makes the framework rigid and would seem to confirm this analysis. However, the alternative method of adding the single segment between $\mathbf{q}$ and $\mathbf{r}$ will also make it rigid! By the counting formula, the framework with the $\mathbf{q}$–$\mathbf{r}$ segment does not have enough segments to be rigid; nevertheless it is.

How can this be?

In this simple example, it is easy to see what is going on: The three segments joining $\mathbf{p}$ to $\mathbf{q}$, $\mathbf{q}$ to $\mathbf{r}$ and $\mathbf{r}$ to $\mathbf{s}$ form a "chain" that is pulled taut, thereby achieving rigidity with fewer than expected edges. If the $\mathbf{q}$-$\mathbf{r}$ segment is added to the right-hand framework, the "chain" hangs loose and the resulting framework is not rigid—as predicted by the counting formula.

Even though this counterintuitive example can be explained, it is very troublesome. It casts some doubt on the entire degrees of freedom approach. Are there other examples of this type that are not so obviously explained?

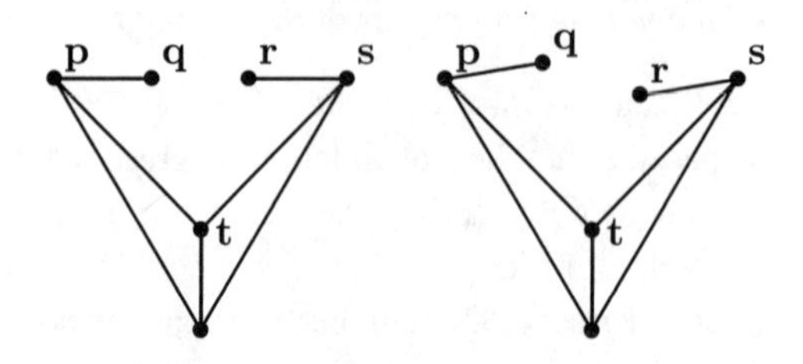

FIGURE 1.17

If so, how can we be sure that a degrees of freedom analysis of a specific framework is valid? Unfortunately, the answer to the first question is yes: There are many not-so-easy-to-explain examples where degrees of freedom arguments incorrectly predict rigidity or nonrigidity. Fortunately, compared to the collection of all frameworks with the same number of points, such examples are extremely rare.

In Chapter 2, we will develop a more precise set of definitions for frameworks and their motions. This second mathematical model of rigidity will enable us to give a complete analysis of 1-dimensional rigidity and to solve the grid problem completely. But this second model (collection of definitions) is difficult to work with in dimension 2 and higher. So, in Chapter 3, we develop a third, more sophisticated, model for rigidity. We will use this third model to justify our simple degrees of freedom model and to explain exactly where it may be safely used. Until then, we will continue to use the degrees of freedom model, as we have developed it, with the understanding that this tool needs refining and that our conclusions must be reviewed when the refinement is accomplished.

The degrees of freedom counting formula for this intuitive approach is easily worked out in any dimension, although we will restrict our discussion to dimensions 1, 2, and 3. The first observation to make is that a point in m-space has m degrees of freedom. Then a framework with n points and s *properly placed* segments should have $mn - s$ degrees of freedom until this number is reduced to the number of degrees of freedom of a rigid body in m-space, df_m. Setting $mn - s = df_m$ and solving for s, we see that we should expect a framework in m-space with n points and $mn - df_m$ *properly placed* segments to be rigid.

We have seen that a rigid body in 2-space, other than a single point, has 3 degrees of freedom, that is, $df_2 = 3$. Furthermore, it is rather clear that any rigid body in 1-space has 1 degree of freedom. Hence, we have only to verify that $df_3 = 6$ in order to justify the third part of the following conjecture.

Conjecture 1.2. *With the exceptions of some very special frameworks such as those discussed above, we expect the following to hold:*

1. *A 1-dimensional framework with n points will need at least $n - 1$ segments to be rigid, and a 1-dimensional framework with n points and $n - 1$ properly placed segments will be rigid.*

2. *A 2-dimensional framework with n points ($n > 1$) will need at least $2n - 3$ segments to be rigid, and a 2-dimensional framework with n points ($n > 1$) and $2n - 3$ properly placed segments will be rigid.*

3. *A 3-dimensional framework with n points ($n > 2$) will need at least $3n - 6$ segments to be rigid, and a 3-dimensional framework with n points ($n > 2$) and $3n - 6$ properly placed segments will be rigid.*

We must verify that a rigid body in 3-space, other than a point or a segment, has 6 degrees of freedom. Consider a triangle:

1. The three joints collectively have 9 degrees of freedom.
2. The 9 degrees are reduced to 6 by the three edges.

We can also look at the triangle in another way:

1. The first vertex has 3 degrees of freedom.
2. Relative to this first vertex, the second vertex can only move on the 2-dimensional surface of a sphere, and it adds 2 more degrees of freedom to the framework.
3. Relative to these first two vertices, the third vertex can only rotate about the line through them, and it adds the sixth degree of freedom to the framework.

For any larger rigid body, we observe that once the three vertices of any embedded triangle are fixed, the entire body is fixed. So any rigid body has 6 degrees of freedom.

The careful reader should have noticed a flaw in the above argument: We made the unwarranted assumption that the three vertices of the triangle *were not collinear*. If you were to randomly select three points in space, the probability that they would be collinear is zero. In other words, having three points in a line is a very special configuration. As we have noted, frameworks in special positions can be exceptions to what is generally true, but we do not yet have mathematical tools powerful enough to account for these exceptions. So, at this stage, we will simply note once again that there may well be exceptions and push on.

Exercise 1.1. *The exact position of a spacecraft is given by six numbers: three position coordinates and three orientation angles (roll, pitch and yaw). Research and describe this 6 degrees of freedom positioning system.*

Exercise 1.2. *Apply this 3-dimensional degrees of freedom approach to the tetrahedron and the cube in Figure 1.1, and the braced cubes in Figure 1.2. In particular check, and perhaps revise, your predictions in Exercise 1.1.1.*

1.5 More About the Grid Problem

We now apply this degrees of freedom analysis to our grid problem. Specifically, we wish to support our conjecture that $h+k-1$ properly placed braces are needed to render the $h \times k$-grid rigid.

Consider the $h \times k$-grid. Let n denote the number of points in this framework. As we have already computed,

$$n = (h+1)(k+1)$$

and s, the number of segments in the grid, is given by

$$s = 2hk + h + k.$$

Assuming that no segment is implied by the others, the grid has

$$\begin{aligned} 2n - s &= 2(h+1)(k+1) - (2hk + h + k) \\ &= h + k + 2 \end{aligned}$$

degrees of freedom. Thus the $h \times k$-grid has

$$(h+k+2) - 3 \qquad \text{or} \qquad h+k-1$$

internal degrees of freedom.

Combining our degrees of freedom approach and our construction approach will yield a useful insight into the deformations of the $h \times k$-grid. Consider attaching a new point by two segments to an existing framework. How does this change the number of internal degrees of freedom?

1. First, adding the new point will increase the number of degrees of freedom by 2.

2. Then, adding each segment will decrease the number of degrees of freedom by 1.

So the net change in the degrees of freedom count is zero.

It follows that removing a point of degree 2 (and its connecting segments) from a framework yields a smaller framework with the same number of degrees of freedom. We also note that by adding a point joined by just one segment to a framework, we produce a larger framework with one more degree of freedom.

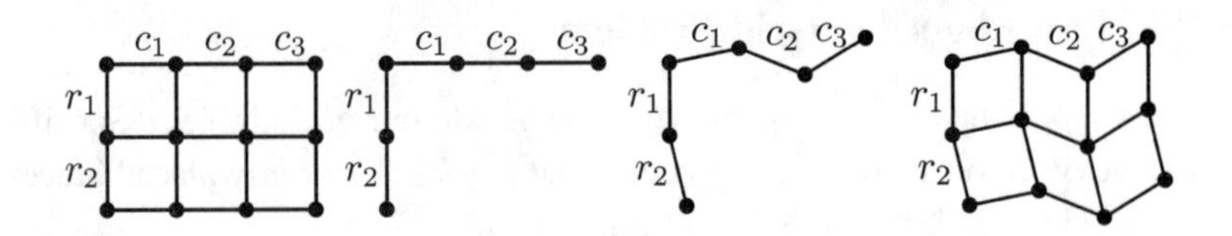

FIGURE 1.18

Now consider the 2×3-grid as pictured on the left in Figure 1.18. It has

$$2 + 3 - 1 = 4$$

internal degrees of freedom, giving it a total of 7 degrees of freedom. Removing points of valence 2, starting with the lower right point and working up, yields the framework consisting of the points and segments along the top and left side of the original framework. We have labeled these segments by r_1, r_2, c_1, c_2 and c_3, denoting the row or column of the original framework that they represent. As we have just observed, this smaller framework also has 7 degrees of freedom. We can identify 3 degrees of freedom of this framework as the 3 degrees of freedom of the rigid framework consisting of r_1 and its endpoints.

Then adding in order each of the segments r_2, c_1, c_2, c_3 along with its other endpoint, we construct the second framework which has an additional 4 internal degrees of freedom. We may exhibit these 4 degrees of freedom by freely choosing the angles that each of these segments makes with the horizontal as we build up the second framework (illustrated by the third framework in the figure). We can then rebuild the entire grid by replacing the valent 2 points in reverse order.

Note that at each step the position of the point is determined—it must complete the rhombus corresponding to the square that was removed along with it. Thus, in general, we can produce all deformations of the $h \times k$-grid by choosing the angles of all but one of the top and left-hand segments—$(h + k - 1)$ choices.

Let's apply this kind of analysis to the two braced grids in Figure 1.14. For the reader's convenience, we reproduce them here in Figure 1.19.

We investigate the right-hand grid in Figure 1.19 first. In Figure 1.20 we start with the (1×3)-grid consisting of the top row of cells. This (1×3)-grid is not rigid and has just one internal degree of freedom which is demonstrated by deforming the last cell. Next, we add the last cell of the second row; since it is braced, it must remain undeformed and can be drawn in only one way.

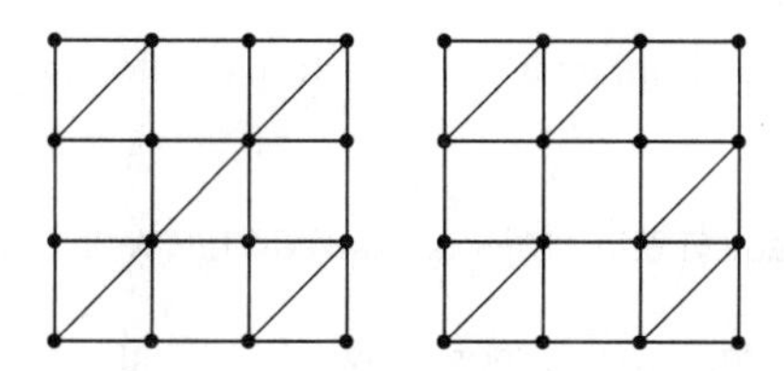

FIGURE 1.19

The rest of the second row is now determined as indicated in the central grid of Figure 1.20. Starting with the first cell of the last row and entering the second cell in the only possible way, we arrive at the lower right-hand cell. This cell, without its brace, can be drawn in only one way. It is not difficult to prove that the four deformed cells are congruent to one another. Hence bracing the last cell in the last row forces each cell to be square; that is, adding this last brace makes the grid rigid.

The first grid in Figure 1.19 can be analyzed in the same manner. But this time, when we get to the last cell, we see that it is forced to be square in spite of any deformation of the rest of the braced grid. Hence the last brace may be included in this deformed grid, demonstrating that this braced grid is not rigid.

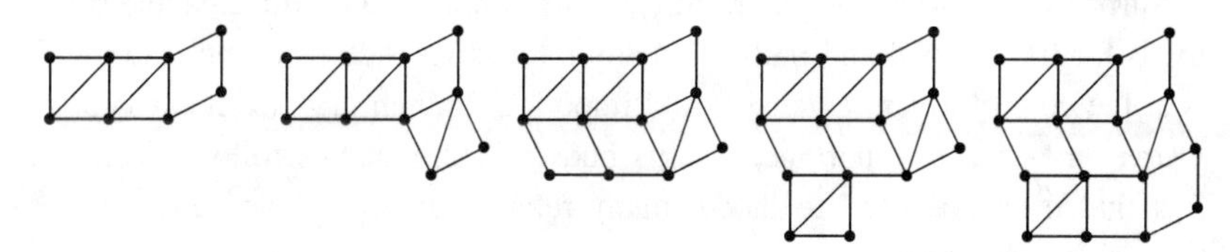

FIGURE 1.20

Exercise 1.1. *Work through this last example step by step. Start with the top row with its central cell deformed; add the (braced) central cell of the second row; fill in the rest of the second row; add the (braced) leftmost cell of the last row and fill in the rest of the third row. Then confirm by a geometric argument that the lower right-hand cell is square.*

Until we develop the additional mathematical tools mentioned above, we can do little more on the theoretical side of the grid bracing problem. However, you can still investigate specific examples. One way to investigate individual problems is to construct physical models. Simple models can be constructed with stiff cardboard strips and thumbtacks; sturdier models can be made with tongue depressors and roofing nails. Another approach is to construct a computer program to simulate a grid. Such a simulation program is available on the web at

<http://www.npac.syr.edu/projects/tutorials/Java/education/grid2/grid3/>

Ultimately we will construct a more powerful "mathematical" model for this problem. This model yields an elegant solution to the general grid problem and will be the culmination of the next chapter.

1.6 What Next?

In the remainder of this book, we will develop and exploit this simple idea of degrees of freedom. Along the way we will completely solve the grid bracing problem. Although plane rigidity (like plane geometry) is a very interesting topic, we live in three dimensions and it is 3-dimensional rigidity that has relevance to real-life architectural and engineering problems. A study of 2-dimensional rigidity may be thought of as a natural step toward a study of 3-dimensional rigidity. Viewing our investigation in this light, it seems natural that the first step should be a study of 1-dimensional rigidity, and, indeed, that is where our in-depth study will start.

Surprisingly, 1-dimensional rigidity, while simpler than 2-dimensional rigidity, is far from trivial and includes many interesting ideas and applications. It is at the core of a very extensive and important branch of mathematics called *graph theory.* The importance of this body of mathematics follows from the fact that it can be used to model many relationships and interdependencies that arise in the real world.

Several such applications are described early in Chapter 2. We will use a graph to model the way a framework is "put together" as opposed to focusing on the actual framework. For example, the four distinct frameworks in Figure 1.21 are all put together in the same way and hence have the same underlying graph or *structure graph*, as we will call it.

As we develop the theory of rigidity, we will see that each framework has two kinds of properties:

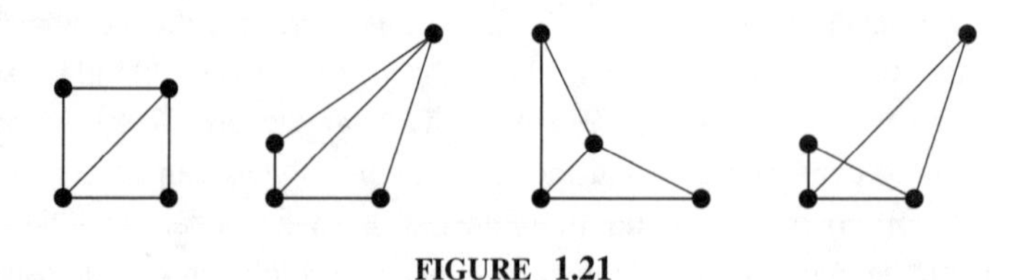

FIGURE 1.21

1. *combinatorial properties*, which are the same for all (or almost all) of the frameworks with the same structure graph; and
2. *geometric properties*, which depend on the positions and dimensions of the rods and joints of the framework.

The next step in our investigation of rigidity is to develop the tools needed to understand the combinatorial properties of frameworks. In Chapter 2, we will explore graph theory and its very close relation to 1-dimensional rigidity. We will then use graph theory in a rather surprising way to solve the grid bracing problem completely.

Chapter 3 is devoted to a development of general rigidity in the plane and in space. There we will see that, unlike 1-dimensional rigidity, 2-dimensional and 3-dimensional rigidity cannot entirely be explained by degrees of freedom arguments. In other words, we will see that 2- and 3-dimensional rigidity are not entirely combinatorial. We will also see that, as of yet, the combinatorial and geometric aspects of a spatial framework cannot be completely separated from one another. In fact, just how to effect such a separation is the subject of present-day mathematical research.

Chapter 4 is devoted to a short exposition of the history of rigidity theory, to investigations of several classes of rod and joint structures in the plane and in 3-space and to an exploration of space structures constructed of rods and wires, held in position by tension.

CHAPTER 2

Graph Theory

2.1 1-Dimensional Rigidity (Part I)

To visualize physical 1-dimensional frameworks, think of the rods and joints as being constrained to move in a glass tube. So, for example, the framework consisting of one long and one short rod joined together can be embedded in 1-space in four different ways, as illustrated in Figure 2.1. (On the actual line, the segments of the folded frameworks would lie on top of one another.)

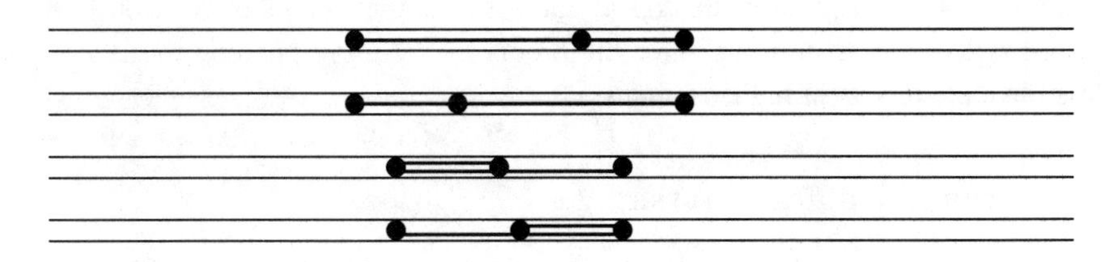

FIGURE 2.1

Clearly, this 1-dimensional framework is rigid, so like all rigid 1-dimensional frameworks it has one degree of freedom: It simply slides along the line but cannot be folded, unfolded or turned around. It is also clear that while a "connected" framework is rigid, one that has more than one piece is not rigid. Those pieces may move independently to give a distortion; see Figure 2.2.

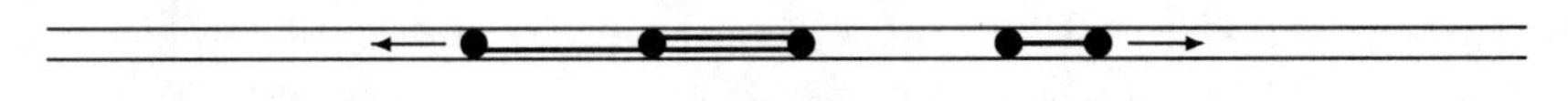

FIGURE 2.2

The degrees-of-freedom approach works very well in dimension one. The number of degrees of freedom of a framework is equal to the number of disjoint pieces of that framework. The framework consisting of k points and no segments has k degrees of freedom. Adding one segment between two of these points gives a framework with $k - 1$ degrees of freedom consisting of $k - 2$ disjoint points and one segment connecting the remaining two points for a total of $k - 1$ pieces. We may now proceed to add segments one at a time. As long as the new segment joins points on different pieces, the number of pieces and the number of degrees of freedom are each reduced by one.

Thus: $k - 1$ properly placed segments will produce a rigid framework on k points.

As we have already noted, the degrees of freedom approach does not always work so easily in dimensions two and higher. So in order to prepare the groundwork for a rigorous development of rigidity in higher dimensions, we will develop 1-dimensional rigidity much more formally than is warranted by an interest in 1-dimensional rigidity alone. In the next section, we give a formal definition of a framework and of the rigidity of a framework both of which are valid in any dimension. In that section, we also state formally (as a theorem) our observation that connectivity and rigidity are equivalent in dimension one. Proving this "obvious" result in this more formal setting is not easy. But the proof of this result and several other proofs of "obvious" results will serve as prototypes for the proofs of the corresponding, but far less obvious, results in higher dimensions.

2.2 Graphs and Frameworks

Recall our intuitive definition of a framework:

> A *planar (linear, spatial) framework* consists of a finite set of points in the plane (line, 3-space), some pairs of which are joined by line segments.

How can we specify which pairs of points are to be joined by segments? A natural way to do this is to label the points and then list the pairs of labels corresponding to the pairs of points to be joined by segments. Another natural way is simply to make a rough sketch of the framework in which the points are not precisely placed and the segments are represented by (freehand) lines or curves joining the points. The purpose of this freehand drawing is simply to record which pairs of points of the framework are to be joined by segments and which are not.

It is natural to use such a freehand drawing in representing any collection of objects, some pairs of which are related while the remaining pairs are not. This simple method of modeling such incidence relations is so powerful that a whole branch of mathematics, called *graph theory*, has grown up around it. In the remainder of this chapter we will develop the basics of graph theory and explore those topics in graph theory that apply to our study of rigidity. Graphs are abstract mathematical constructions that are used to model simple relationship structures from the real world.

For example, suppose we let W denote the five workers in a small carpentry shop that makes table legs, $W = \{a, b, c, d, e\}$ for Al, Betty, Cliff, Dan and Ethel; and suppose we let S denote the four skills needed in the process of making table legs, $S = \{f, s_1, s_2, t\}$ for finishing, sawing, sanding and turning (on a lathe). We can model the relationship between workers and tasks by making a list of pairs consisting of a worker and a task he or she can perform. For instance, assume that

$$E=\{\{a,f\},\{a,s_1\},\{b,s_1\},\{c,s_1\},\{c,s_2\},\{c,t\},\{d,s_2\},\{e,s_1\},\{e,t\}\}$$

is a complete listing of such pairs. The two sets $V = W \cup S$ and E encode all of the information about "who can do what" in our little carpentry shop. To get a "picture" of this information we select nine points in the plane, label them by the elements of V and draw in a segment for each pair in E. This yields the "graph" in Figure 2.3.

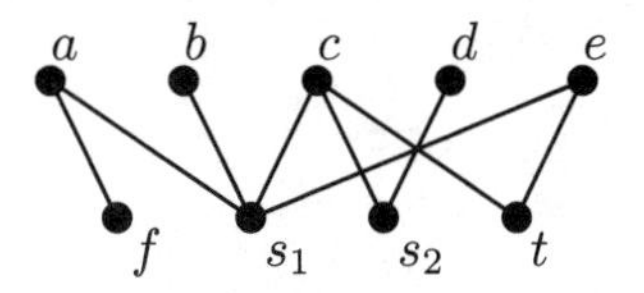

FIGURE 2.3

From this graph, we easily see that Dan can do only one job, sanding; that everyone except Dan can saw the rough pieces from which the legs are made; that only Cliff and Ethel can run the lathe; and so on. Because of this ease of use, graphs are almost always presented as diagrams rather than lists of pairs.

More formally:

> A *graph* is a pair (V, E), where V is a finite set whose elements are the *vertices* of the graph and E is a collection of pairs of vertices called the *edges* of the graph.

If a and b are vertices ($a, b \in V$) and the pair $e = \{a, b\}$ is an edge ($e \in E$), then a and b are called the *endpoints* of the edge e, and a and b are said to be *adjacent.* Note that an edge consists only of a pair of vertices; however, when we draw a picture of a graph, we draw a curve or line segment joining the endpoints of the edge. Since these curves only indicate which pairs are edges, just how they are drawn has no significance. In particular, no significance is given to any points at which these curves or segments may cross.

Take as another example of a graph the "acquaintance graph" (V, E) of a group of people. The vertices of the graph denote the people in the group, and an edge $\{a, b\}$ is included if the people represented by a and b are acquainted. Such an acquaintance graph is diagrammed in Figure 2.4. The acquaintances among these eight people are easily understood and many features of these relationships are easily recognized using this graph. For example s, u, v and w represent a group of four people who are mutually acquainted. Furthermore, it is easy to see that there is no other group of four and no group of more than four who are mutual acquaintances. The people represented by t, u and x form a group of three mutual strangers.

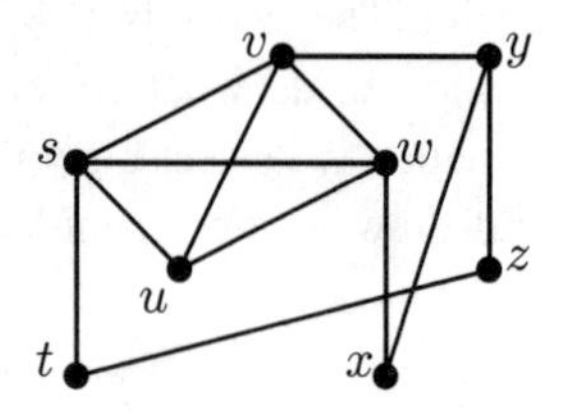

FIGURE 2.4

Exercise 2.1. *Refer to Figure 2.4.*

1. *Show that there is no set of four mutual strangers.* [Hint: Look at the smaller graph with vertices $\{s, u, v, w\}$ and the smaller graph with vertices $\{x, y, z, t\}$.]

2. *Find all eight different groups of three mutual strangers.*

Using acquaintance graphs, one may solve a problem that frequently appears as a math puzzle:

> *Show that any group of six people must include either a group of three mutual acquaintances or a group of three mutual strangers.*

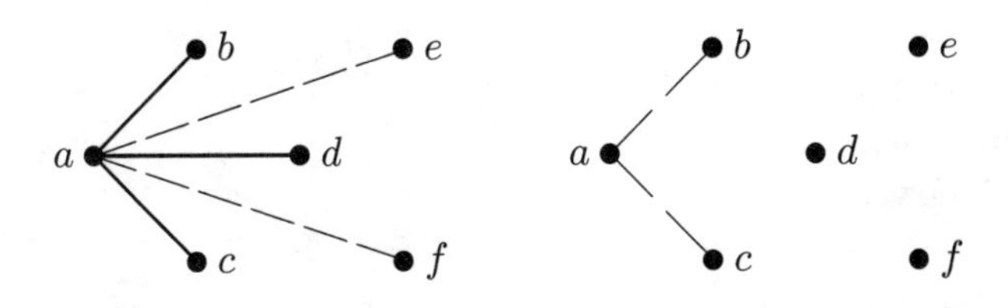

FIGURE 2.5

Assume we have an acquaintance graph for six people and denote the vertices by a, b, c, d, e and f. Suppose a is acquainted with three (or more) of the others; let's say a is acquainted with b, c and d and possibly e or e and f. See the left-hand diagram in Figure 2.5. If b and c are acquainted, a, b and c is a group of three mutual acquaintances. Similarly, we have a group three mutual acquaintances if b and d are acquainted or if c and d are acquainted.

On the other hand, if none of the pairs $\{b, c\}$, $\{b, d\}$ and $\{c, d\}$ is acquainted, b, c and d form a group of three mutual strangers. We conclude that, if a is acquainted with three (or more) of the others, then there must be either a group of three mutual acquaintances or a group of three mutual strangers.

Next suppose that a is acquainted with two (or fewer) of the others—say none, or b, or b and c. See the right-hand diagram in Figure 2.5. Then a is not acquainted with d, e and f. If any two of d, e and f are not acquainted, those two and a form a group of three mutual strangers; if on the other hand, each pair is acquainted, we have a group of three mutual acquaintances. Thus, in any case, either there is a group of three mutual acquaintances or a group of three mutual strangers.

Exercise 2.2. *Construct a graph that shows that it is possible for a group of five people to contain no group of three mutual acquaintances and no group of three mutual strangers.*

Another simple but surprising result about acquaintances may be verified using an acquaintance graph. It is left as an exercise for the interested reader.

Exercise 2.3. *In any group of individuals, the number of individuals in the group who are acquainted with an odd number of other individuals in the group is even.* [Hint: Sum over all individuals the number of his or her acquaintances and relate this sum to the number of edges in the acquaintance graph.]

One of the very first applications of graph theory was to chemistry, specifically to the structure of molecules. The atoms are represented by the vertices

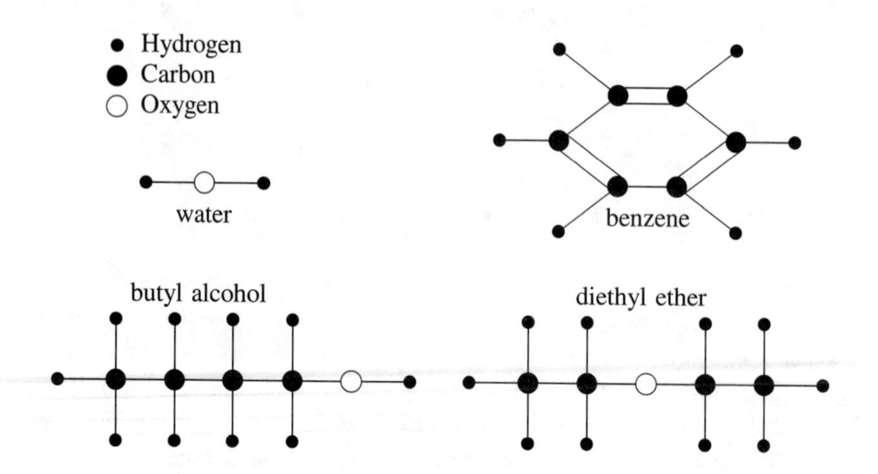

FIGURE 2.6

of a graph and the chemical bonds by edges. In Figure 2.6, we picture a molecule of each of four compounds: water, H_2O; benzene, C_6H_6; and two other compounds that are very interesting in that they have the same molecular formula, $C_4H_{10}O$, but have different chemical properties. The reason for these differing chemical properties is that they are "connected up" differently. This is easily seen from the graph model. In fact, the graph is intended to show only how the atoms of the molecule connect up. A physical model of a molecule contains much more information and is, of course, much more complicated. To start with, a physical model of a molecule would have to be 3-dimensional and it would have to have precise angle and distance measurements.

The double edges in the benzene ring indicate a double chemical bond. Graphs that permit multiple edges are called *multigraphs*. Multigraphs have many applications—but rigidity theory is not one of them. So we will not consider multigraphs, even though much of our development of graph theory extends naturally to multigraphs.

Just as the chemists use graphs to diagram how molecules are put together, we will use graphs to diagram how frameworks are put together. We are now ready to give a formal definition of a framework in m-dimensional space:

> An *m-dimensional framework* $(V, E, \mathbf{p})$ consists of a graph (V, E) and a function $\mathbf{p}$ from the vertex set into m space, $\mathbf{p} : V \to R^m$.

The definition is designed for mathematical precision and ease in proving mathematical results; to link this definition to our intuitive concept of a framework, think of the points

$$\mathbf{p}(a), \qquad a \in V,$$

as the joints of the framework, and think of the segments

$$\overline{\mathbf{p}(a)\mathbf{p}(b)}, \qquad \{a, b\} \in E,$$

as the rods. The graph (V, E) is the diagram that shows how the framework is put together. We will be concerned only with the cases $m = 1, 2, 3$, but this and the definitions to follow work in all higher dimensions as well.

For convenience, we introduce some terminology:

- The graph (V, E) is called the *structure graph* of the framework $(V, E, \mathbf{p})$.
- The function $\mathbf{p}$ is called an *embedding* of V into m-space.

Since frameworks are rather complicated mathematical structures, it is important to adopt a system of notation that helps us to keep track of the various objects. Frequently we will index the vertices of the structure graph, $V = \{a_1, \dots, a_n\}$ and when we do:

- The point $\mathbf{p}(a_i)$ is denoted by $\mathbf{p}_i$.
- Its coordinates are denoted by

$$(x_i), \quad (x_i, y_i) \quad \text{or} \quad (x_i, y_i, z_i),$$

in 1-, 2- or 3-dimensional space, respectively.

Properties of a framework that depend only on how the rods are connected together and not on the positions of the joints or the lengths of the rods should be discernible directly from the structure graph of that framework and should be shared by all other frameworks with the same structure graph. As we have already noted, we call these the *combinatorial properties* of the framework. In other words, the combinatorial properties of a framework are those that do not depend on the embedding function. The remaining properties, that is, those that do depend on the embedding function, are called the *geometric properties* of the framework.

Now comes the big question:

Is rigidity a combinatorial property?

It follows from our informal discussions (to be formally proved later on) that the answer is *yes* in dimension one. Sorting out the combinatorial and geometric properties of frameworks in dimensions two and three is one of the two main themes of this book. The other is how to take advantage of the combinatorial properties in designing and constructing frameworks.

Let's consider an example. Take the first framework we considered, the cube (Figure 1.1). In Figure 2.7, we draw its structure graph. If we coordinatize 3-space so that our cube is the unit cube in the first octant, its formal description as a framework is given by $(V, E, \mathbf{p})$ where:

- (V, E) is the structure graph with vertices labeled $\{a_1, \ldots, a_n\}$, as pictured in Figure 2.7, and where
- the mapping $\mathbf{p}$ is defined by

$$\begin{array}{llll} \mathbf{p}_1 = (0,0,0), & \mathbf{p}_2 = (0,1,0), & \mathbf{p}_3 = (1,1,0), & \mathbf{p}_4 = (1,0,0), \\ \mathbf{p}_5 = (0,0,1), & \mathbf{p}_6 = (0,1,1), & \mathbf{p}_7 = (1,1,1), & \mathbf{p}_8 = (1,0,1). \end{array}$$

This model tallies with the picture in Chapter 1 if we think of the positive x axis as pointing to the right, the positive z direction as up and the positive y direction as into the page.

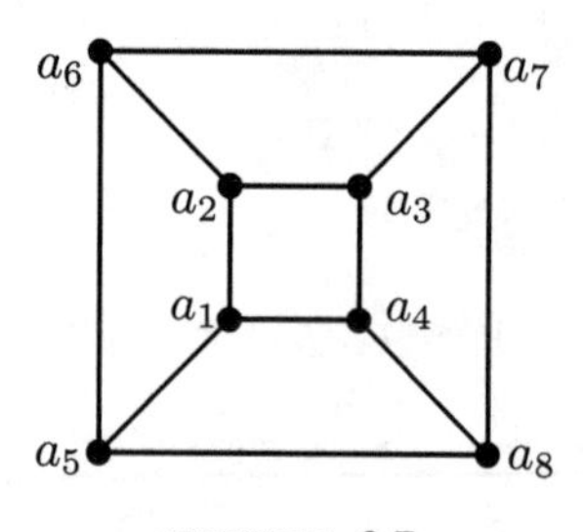

FIGURE 2.7

Exercise 2.4. *Give a formal description for each of the braced cubes pictured in Figure 1.2. Use the same labeled vertex set and the same embedding function* $\mathbf{p}$ *as used above for the cube; that is, simply add the appropriate edges to the structure graph of the unbraced cube. You may wish to use some curved arcs to avoid edges being drawn on top of one another.*

This definition of a framework allows us to separate the way that the framework is connected up from the actual positions of the joints and rods in m-space. We illustrate this in Figure 2.8 with two additional embeddings of

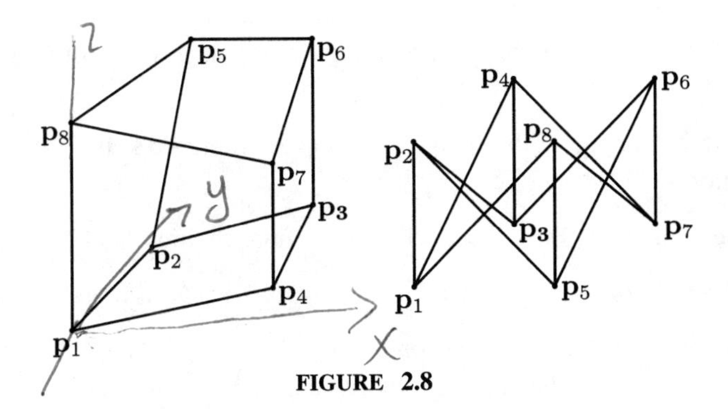

FIGURE 2.8

the graph in Figure 2.7. The framework on the left is easily recognized as a distorted cube; its embedding function is

$$\begin{aligned} &\mathbf{p}_1 = (0,0,0), \quad &&\mathbf{p}_2 = (0,1,0), \quad &&\mathbf{p}_3 = (1,1,0.3), \\ &\mathbf{p}_4 = (1,0,0.1), \quad &&\mathbf{p}_5 = (0.1,1.2,1.1), \quad &&\mathbf{p}_6 = (1,1,1.2), \\ &\mathbf{p}_7 = (1,0,0.9), \quad &&\mathbf{p}_8 = (0,0,1.2). \end{aligned}$$

On the other hand, the embedding pictured on the right is not at all recognizable as a cube; its embedding function is

$$\begin{aligned} &\mathbf{p}_1 = (0,0,0), \quad &&\mathbf{p}_2 = (0,0,1), \quad &&\mathbf{p}_3 = (0,1,0), \quad &&\mathbf{p}_4 = (0,1,1), \\ &\mathbf{p}_5 = (1,0,0), \quad &&\mathbf{p}_6 = (1,1,1), \quad &&\mathbf{p}_7 = (1,1,0), \quad &&\mathbf{p}_8 = (1,0,1). \end{aligned}$$

Seeing these two very different frameworks with the same structure graph as the cube makes it clear that the question "Is rigidity a combinatorial property?" will not be an easy one to answer. In fact, based on these examples, one may begin to doubt that rigidity is a combinatorial property!

But consider this: No matter how the tetrahedral graph is embedded, the resulting framework is a tetrahedron or a collapsed tetrahedron and, in either case, is rigid. Hence, if we add the braces pictured on the left in Figure 1.2 to the graph in Figure 2.7, then add the corresponding segments to the frameworks in Figure 2.8, those frameworks, like the braced cube on the left in Figure 1.2, will consist of four tetrahedra glued together and will therefore be rigid.

Exercise 2.5. *Carry out the constructions just described and verify the conclusions.*

To define the concept of a motion of a framework so that it also fits our intuition takes some care. It is natural to think of a motion as taking place over time. So let the variable t, representing time, range from 0 to 1 on the real line, $t \in [0, 1]$:

- At the beginning of the motion, $t = 0$, the framework is in its initial position;
- at the end, $t = 1$, the framework is in its final position; and
- the framework is in transitional positions for t between 0 and 1.

We also want the change in position of each vertex to be a "nice" function of time. We will all agree that, for physical reasons, motions must be continuous. But we will also insist that the motions be "smooth," that is, differentiable. This condition gives us a model with nicer motion functions but rules out the jerky motions that seem physically possible.

The mathematical model builder is frequently in this situation: wishing to make a model that accurately reflects the reality being modeled, yet wishing to make the model easy to work with. We can justify our choice by arguing that, if there is a jerky motion that deforms a framework, then there must also be a smooth motion that deforms it. This seems reasonable and, if it is true, the restriction to smooth motions will not alter the classification of frameworks as rigid or nonrigid.

We should point out that we have already made another modeler's choice: We did not require that the embedding function $\mathbf{p}$ for a framework be one to one. It certainly seems natural to do so. It is hard to imagine the need for permitting a framework to have two distinct joints occupying the same position.

But consider the square in the plane. We can deform it into a rhombus and, if we continue the motion, opposite vertices will eventually pass one another. If we require the embedding function to be one to one, the deforming square would cease to be a framework for that instant when the two vertices coincide. On the other hand, we may actually wish to avoid this case, since strange things can happen when the vertices do coincide. At that point, the two vertices could stay together and the flattened square could begin to fold into an "L" shape.

For now we will stick with simplest (mathematically speaking) choices:

- We will put no conditions on the embedding functions; and
- we will only require the motion functions to satisfy a simple differentiablity condition.

Later, we may reconsider these choices.

We formalize our definition of a motion of a framework as follows. Let $(V, E, \mathbf{p})$ be a framework with indexed vertex set $V = \{a_1, \ldots, a_n\}$. Then:

- A *motion* of this framework comprises an indexed family of functions $\mathbf{P}_i : [0,1] \to R^m$, $i = 1, \ldots, n$, so that:
  1. $\mathbf{P}_i(0) = \mathbf{p}_i$, for all i;
  2. $\mathbf{P}_i(t)$ is differentiable on the interval $[0,1]$, for all i;
  3. $|\mathbf{P}_i(t), \mathbf{P}_j(t)| = |\mathbf{p}_i, \mathbf{p}_j|$, for all $t \in [0,1]$ and $\{a_i, a_j\} \in E$.
- The function $\mathbf{P}_i(t)$ is called the *trajectory* of the point $\mathbf{p}_i$ under the motion.
- The notation $|\mathbf{p}_i, \mathbf{p}_j|$ denotes the distance between the points $\mathbf{p}_i$ and $\mathbf{p}_j$ in R^m. We have

$$|\mathbf{p}_i, \mathbf{p}_j| = |x_i - x_j| = \sqrt{(x_i - x_j)^2}, \qquad \text{if } m = 1;$$

$$|\mathbf{p}_i, \mathbf{p}_j| = \sqrt{(x_i - x_j)^2 + (y_i - y_j)^2}, \qquad \text{if } m = 2;$$

$$|\mathbf{p}_i, \mathbf{p}_j| = \sqrt{(x_i - x_j)^2 + (y_i - y_j)^2 + (z_i - z_j)^2}, \qquad \text{if } m = 3.$$

- For a fixed time t, the framework $(V, E, \mathbf{q})$, where $\mathbf{q}_i = \mathbf{P}_i(t)$, is the position to which the the initial framework has moved at time t.
- A motion $\{\mathbf{P}_i\}$ of the framework $(V, E, \mathbf{p})$ is a *rigid motion* if all distances between vertices are preserved by the motion:

$$|\mathbf{P}_i(t), \mathbf{P}_j(t)| = |\mathbf{p}_i, \mathbf{p}_j|,$$

  for all $t \in [0,1]$ and all $1 \leq i < j \leq n$.
- A motion $\mathbf{P}$ of the framework $(V, E, \mathbf{p})$ is a *deformation* if the distance between at least one pair of vertices is changed by the motion:

$$|\mathbf{P}_i(t), \mathbf{P}_j(t)| \neq |\mathbf{p}_i, \mathbf{p}_j|,$$

  for some $t \in [0,1]$ and some $\{a_i, a_j\} \notin E$.
- A framework $(V, E, \mathbf{p})$ is said to be *rigid* if all of its motions are rigid motions, that is, if it admits no deformations.

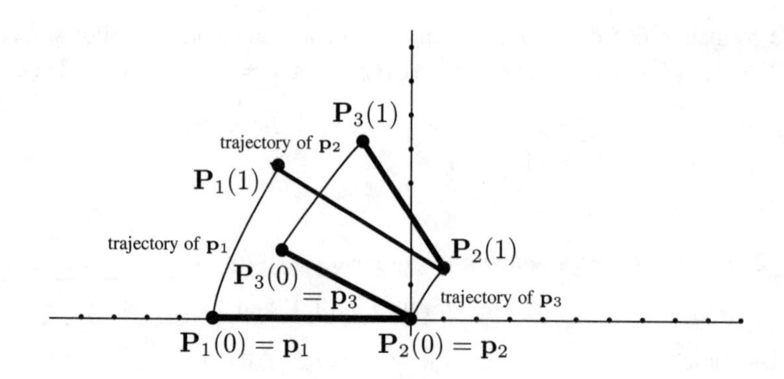

FIGURE 2.9

At this point it may be worth the effort to work through the details of a simple example. Let $V = \{a_1, a_2, a_3\}$, $E = \big\{\{a_1, a_2\}, \{a_2, a_3\}\big\}$ and let $\mathbf{p} : V \to R^2$ be defined by

$$\mathbf{p}(a_1) = \mathbf{p}_1 = (-6, 0),$$

$$\mathbf{p}(a_2) = \mathbf{p}_2 = (0, 0),$$

$$\mathbf{p}(a_3) = \mathbf{p}_3 = (-4, 2).$$

We depict this framework $\mathcal{F} = (V, E, \mathbf{p})$ in Figure 2.9.

Now consider the functions:

$$\mathbf{P}_1(t) = \left(t - 3\sqrt{4 - t^2},\, 5t - \frac{t^2}{2}\right);$$

$$\mathbf{P}_2(t) = \left(t,\, 2t - \frac{t^2}{2}\right);$$

$$\mathbf{P}_3(t) = \left(2t - 2\sqrt{4 - t^2},\, 4t - \frac{t^2}{2} + \sqrt{4 - t^2}\right).$$

One verifies that these functions describe a motion of $\mathcal{F}$ by computing:

$$|\mathbf{P}_1(t), \mathbf{P}_2(t)| = \sqrt{\left(t - 3\sqrt{4 - t^2} - t\right)^2 + \left[5t - \frac{t^2}{2} - \left(2t - \frac{t^2}{2}\right)\right]^2}$$

$$= \sqrt{\left(-3\sqrt{4 - t^2}\right)^2 + (3t)^2} = 6, \qquad \text{for all } t,$$

and computing $|\mathbf{P}_3(t), \mathbf{P}_2(t)|$ to be $2\sqrt{5}$, for all t. We then show that this motion is a rigid motion, as illustrated in Figure 2.9, by showing that

$$|\mathbf{P}_3(t), \mathbf{P}_1(t)| = 2\sqrt{2}, \qquad \text{for all } t.$$

Exercise 2.6. *Complete the proof that these functions are a motion of $\mathcal{F}$ by showing that*

$$|\mathbf{P}_3(t), \mathbf{P}_2(t)| = 2\sqrt{5}, \qquad \text{for all } t.$$

Demonstrate that it is, in fact, a rigid motion by showing that

$$|\mathbf{P}_3(t), \mathbf{P}_1(t)| = 2\sqrt{2}, \qquad \text{for all } t.$$

If we change the third function $\mathbf{P}_3(t)$ to

$$\mathbf{P}_3(t) = \left(t\sqrt{4-t^2} + 2t^2 + t - 4,\ 2t\sqrt{4-t^2} - \frac{3t^2}{2} + 2t + 2\right)$$

while retaining $\mathbf{P}_1(t)$ and $\mathbf{P}_2(t)$ unchanged, we get another motion of $\mathcal{F}$. But this one is a deformation; see Figure 2.10.

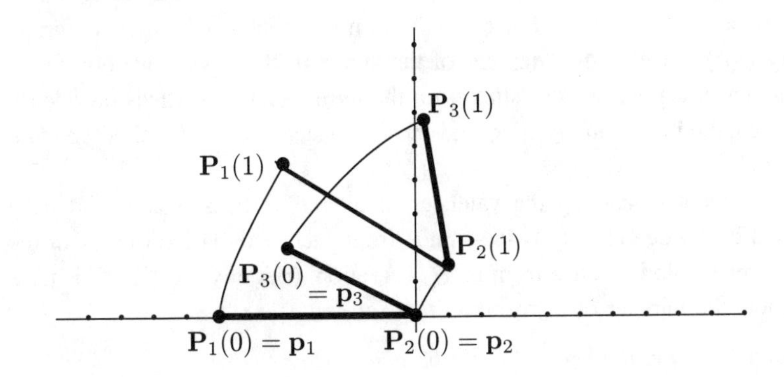

FIGURE 2.10

Exercise 2.7. *Show that $\mathbf{P}_1(t)$ and $\mathbf{P}_2(t)$ along with the new $\mathbf{P}_3$ are a motion of $\mathcal{F}$. Note that we still have that $|\mathbf{P}_1(t), \mathbf{P}_2(t)| = 6$, and show that $|\mathbf{P}_3(t), \mathbf{P}_2(t)| = 2\sqrt{5}$, for all t. Next show that they are, in fact, a deformation by showing that $|\mathbf{P}_3(t), \mathbf{P}_1(t)| \neq |\mathbf{p}_3, \mathbf{p}_1|$, for some t.*

With the definitions of frameworks, motions and rigidity well in hand, we return to our consideration of 1-dimensional rigidity. Our informal discussion in the previous section led us to the observation that 1-dimensional rigidity corresponds to connectivity of its underlying graph and is therefore a combinatorial property. Before we can give a formal proof of this correspondence, we must give a formal definition of connectivity in graphs; this is done in the next section. Graph theory is an extensive subject. In the next few sections, we barely scratch the surface as we develop the few topics from graph theory necessary for our study of rigidity.

2.3 Fundamentals of Graph Theory

Since connectivity in graphs is of primary interest to us, we start with a formal definition of that concept. A graph (V, E) is said to be *disconnected* if the vertex set can be partitioned into two nonempty sets A and B ($V = A \cup B$, $A \cap B = \emptyset$, $A \neq \emptyset$ and $B \neq \emptyset$) so that no edge has one endpoint in A and the other endpoint in B. We say that a graph (V, E) is *connected* if no such partition exists.

We now turn to some of the other basic terminology of graph theory. Let (V, E) be a graph and let $a \in V$. The number of edges of the graph with the vertex a as an endpoint is called the *valence* of a and is denoted by $\rho(a)$. One of the unfortunate features of graph theory is that there is no universal agreement on terminology. For example, in many books and papers on graph theory, $\rho(a)$ is called the "degree" of the vertex a. Since we have other uses for the term degree, we will stick with the term valence. Vertices of valence zero are called *isolated vertices,* and those of valence one are called *pendant vertices.*

If you were to add up the valences of all the vertices, you would have counted each edge exactly twice—once from each end. The formulas in the next lemma follow at once from this observation. (For any set X, $|X|$ is used to denote the number of elements in that set.)

Lemma 2.1. *Let the graph* (V, E) *be given. Then*

$$\sum_{x \in V} \rho(x) = 2|E|,$$

and the average valence in (V, E) *is*

$$\overline{\rho} = \frac{2|E|}{|V|}.$$

Several classes of graphs occur so frequently that they have been assigned special names. First is the class of *complete graphs:*

- A graph (V, E) with $|V| = n$ and E consisting of all

$$\binom{n}{2} = \frac{n(n-1)}{2}$$

pairs of vertices is called a *complete* graph.

Clearly, two complete graphs on the same number of vertices have the same structure as graphs; in particular, they can be represented by the same picture. To be precise:

- Two graphs (V, E) and (U, D) are *isomorphic* if there is a one-to-one correspondence $c : V \to U$ so that $\big(c(a_i), c(a_j)\big) \in D$ if and only if $(a_i, a_j) \in E$.

We tend to group together all isomorphic copies of a graph and speak of them as one graph. In that sense, we speak of "the" complete graph on n vertices and denote "it" by K_n. In Figure 2.11, we draw the first five complete graphs.

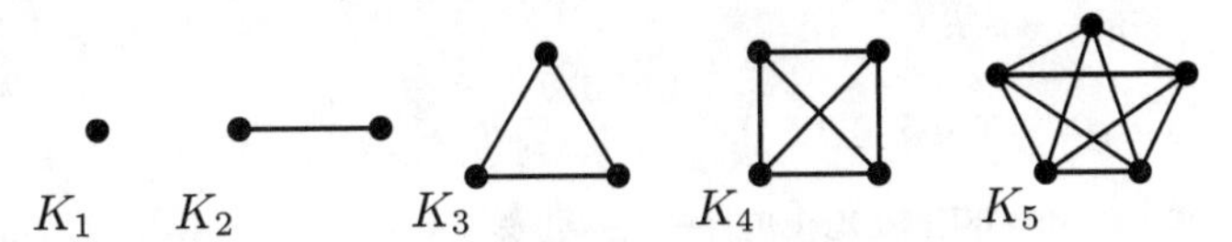

FIGURE 2.11

Continuing with our terminology, again let (V, E) be a graph.

- Any graph (U, D) consisting of some vertices and some edges of (V, E) ($U \subseteq V$ and $D \subseteq E$) is called a *subgraph* of (V, E).
- If $U = V$, then $(U, D) = (V, D)$ is called a *spanning subgraph* of (V, E).

For example, every graph on n vertices is a spanning subgraph of the complete graph on the same vertex set.

- A graph (V, E) is said to be *bipartite* if it has a partition of its vertex set into two cells, A and B ($V = A \cup B$, $A \cap B = \emptyset$), so that every edge in E has one endpoint in each cell.

The graph in Figure 2.3 is bipartite; the graph in Figure 2.4 is not.

If $|A| = m$, $|B| = n$ and E is the set of all $m \times n$ pairs of vertices with one from A and one from B, then $(A \cup B, E)$ is called a *complete bipartite* graph; "the" complete bipartite graph with "parts" of size m and n is denoted by $K_{m,n}$. We draw seven of the smallest complete bipartite graphs in Figure 2.12.

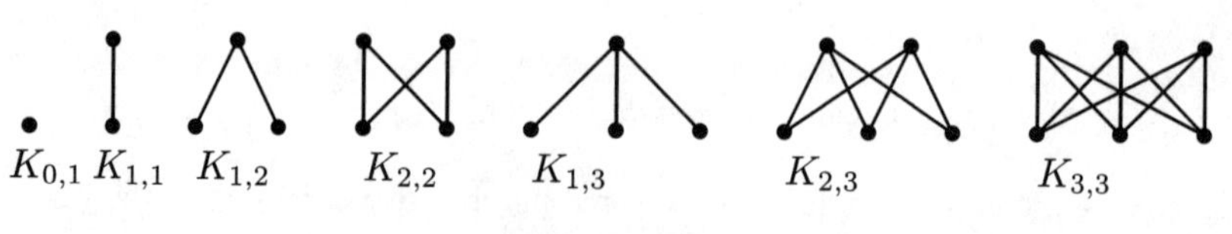

FIGURE 2.12

Exercise 2.1. *Show that a bipartite graph on n vertices has average valence at most $n/2$.*

Still continuing with our terminology, let (V, E) be a graph and let a and b be vertices. By a *path* from a to b, we mean a sequence of vertices $a_0, a_1, \ldots, a_k$ so that:

1. $a = a_0$ and $a_k = b$;
2. $a_0, a_1, \ldots, a_k$ are distinct;
3. a_{i-1} and a_i are adjacent, for $i = 1, \ldots, k$.

The edges joining successive vertices in the sequence are called the *edges of the path*, and the number of these edges is called the *length* of the path. The graph that consists entirely of the vertices and edges of a path of length n is denoted by P_n. Note that

$$P_0 = K_1 = K_{0,1}, \quad P_1 = K_2 = K_{1,1} \quad \text{and} \quad P_2 = K_{1,2}.$$

Assume the graph (V, E) has a sequence of vertices $a = a_0, a_1, \ldots, a_k = b$ so that each successive pair of vertices in the sequence is joined by an edge but that the vertices are not necessarily distinct. We may think of such a sequence as a meandering "walk" through the graph that may crisscross or even backtrack on itself.

Now suppose that $a_i = a_j$, for $i < j$; then we may construct a shorter sequence joining a to b by simply deleting the vertices a_{i+1} through a_j and re-indexing the vertices beyond a_j. If we repeat this construction as often as possible, the final result will be a path from a to b in the graph. We reformulate this into a formal statement to which we can easily refer:

Lemma 2.2. *Let the graph (V, E) be given. If there is a sequence of vertices*

$$a = a_0,\ a_1, \ldots,\ a_k = b$$

so that every successive pair of vertices in the sequence is joined by an edge, then a and b are joined by a path in the graph.

One useful result that follows directly from this lemma is:

Lemma 2.3. *Given a graph (V, E), if it admits a path from a to b and another path from b to c, then it admits a path from a to c.*

To see this, note that the path from a to b followed by the path from b to c is a sequence of vertices starting with a and ending with c so that every successive pair of vertices in the sequence is joined by an edge. Hence, by Lemma 2.2, a and c are joined by a path in the graph. We may now give a useful characterization of connectivity in terms of paths.

Lemma 2.4. *A graph (V, E) is connected if and only if every pair of its vertices is joined by a path.*

Proof. Assume that there are two vertices a and b in V that are not joined by a path. Let A be the set of all vertices c so that there is a path from a to c and let $B = V - A$. Since $a \in A$ and $b \in B$, A and B partition V into two nonempty sets. Suppose that there is an edge $e = \{a', b'\}$ with $a' \in A$ and $b' \in B$. Then by Lemma 2.3, the path from a to a' and the path (of length 1) from a' to b' yield a path from a to b'. But, this contradicts the fact that b' is not in A. We conclude that there is no edge with one endpoint in A and the other in B; hence (V, E) is not connected.

Now assume that every two vertices of (V, E) are joined by a path and let A and B be any partition of V into two nonempty cells. Choose $a \in A$ and $b \in B$ and let $a = a_0,\ a_1, \ldots,\ a_k = b$ be a path joining a and b. Let a_i be the last vertex along the path in A; since $b \in B$, $i < k$. Then the next vertex a_{i+1} is in B and the edge $\{a_i, a_{i+1}\}$ has one endpoint in A and the other in B. We conclude that (V, E) is connected. □

Lemma 2.5. *Let a graph (V, E) be given. Then there are connected graphs $(V_1, E_1), \ldots, (V_k, E_k)$ so that $V_1, \ldots V_k$ are disjoint, $V = V_1 \cup \cdots \cup V_k$ and $E = E_1 \cup \cdots \cup E_k$.*

Proof. If (V, E) is connected, take $k = 1$, $V_1 = V$ and $E_1 = E$. Assume then that (V, E) is not connected. Select a vertex $a_1 \in V$ and let $V_1 = \{a : a \textit{ is joined by a path to } a_1\}$. If $(a, b) \in E$ and $a \in V_1$ then, by Lemma 2.3, there is a path from a_1 to b. We conclude that any edge with one endpoint in V_1 has both endpoints in V_1. Let E_1 be the set of all edges with both endpoints in V_1. Thus, all edges that lie on paths between vertices in V_1 belong to E_1, and (V_1, E_1) is a connected subgraph of (V, E).

Next note that each edge in $E - E_1$ has both of its endpoints in $V - V_1$. Hence (V, E) decomposes into the two disjoint graphs (V_1, E_1) and $(V - V_1, E - E_1)$. If $(V - V_1, E - E_1)$ is connected, we are done; if not, apply the above construction to further decompose $(V - V_1, E - E_1)$. Since V is finite, we can repeat the construction only a finite number of times and the final result is the decomposition we seek. □

The connected graphs in the decomposition of (V, E) are called the *connected components* (or simply *components*) of (V, E). By a *circuit* in a graph, we mean a sequence of vertices $a_0, a_1, \ldots, a_k$ so that:

1. $a_0 = a_k$;
2. $a_1, \ldots, a_k$ are distinct;
3. a_{i-1} and a_i are adjacent, for $i = 1, \ldots, k$.

The edges joining successive vertices in the sequence are called the *edges of the circuit* and the number of these edges is called the *length* of the circuit. The graph that consists entirely of the vertices and edges of a circuit of length n is denoted by C_n. Note that $C_3 = K_3$ and $C_4 = K_{2,2}$.

2.4 Trees

One of the most important classes of graphs for the study of rigidity is the class of *trees*:

- A graph that is connected and contains no circuit is called a *tree.*

Figure 2.13 pictures all trees on five or fewer vertices.

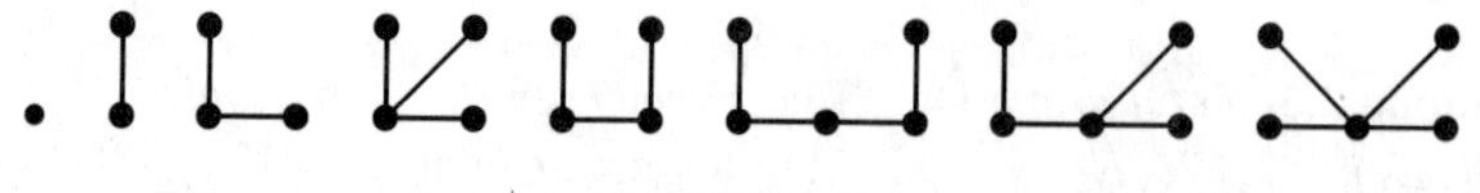

FIGURE 2.13

Exercise 2.1.

1. *Draw all six trees on six vertices.*

2. *Draw all eleven trees on seven vertices.*

A graph that is circuit-free may not be connected and therefore not a tree. However, each of its components is connected and circuit-free; that is, each component is a tree. So, a graph that has no circuits is either a tree or a *forest:*

- A graph whose connected components are trees is called a *forest.*

Many of the proofs in this section will be inductive, so we will be interested in what happens when we delete an edge from a graph, particularly from a connected graph. When $e \in E$, we use the notation $E - e$ to denote the set obtained by deleting e from E.

Lemma 2.6. *Let (V, E) be a connected graph and $e \in E$ an edge. Then $(V, E - e)$ is connected if and only if e lies on a circuit in (V, E).*

Proof. First, assume that $e = \{a, b\}$ belongs to a circuit. Let c and d be any two vertices in V. Since (V, E) is connected, there is a path from c to d in (V, E). If that path does not "contain" the edge e, it is also a path in $(V, E-e)$. Suppose then that it does contain the edge e; without loss of generality we may assume that this path has the form $c = c_1, \ldots, a, b, \ldots, c_k = d$. Since e is on a circuit in (V, E), a and b are still joined by a path in $(V, E - e)$. We have then, in $(V, E - e)$, a path from c to a, a path from a to b and a path from b to d; by Lemma 2.2, $(V, E - e)$ admits a path from c to d. We conclude that $(V, E - e)$ is connected.

Next assume that $e = \{a, b\}$ and that $(V, E - e)$ is connected. By Lemma 2.4, there is a path $a = a_1, \ldots, a_k = b$ in $(V, E - e)$ from a to b. But then $a = a_1, \ldots, a_k = b, a$ is a circuit in (V, E) containing e. □

The next result relates trees to our central interest, connectivity.

Lemma 2.7. *A graph (V, E) is connected if and only if it has a spanning tree (a spanning subgraph that is a tree).*

Proof. Assume that (V, T) is a spanning tree of (V, E) and let $a, b \in V$. Since (V, T) is connected, it admits a path from a to b and, since $T \subseteq E$, this path is also a path in (V, E). Hence (V, E) is connected.

Now, assume that (V, E) is connected. If (V, E) is not a tree, it must contain a circuit. Let $E' = E - e$ where e lies on a circuit in (V, E). Then

(V, E') is a spanning subgraph of (V, E) and, by the previous lemma, it is still connected. If (V, E') is not a tree, we may delete another edge from a circuit of (V, E') to get (V, E''), another connected spanning subgraph of (V, E) with even fewer edges. Clearly this deletion process must eventually stop and it can only stop with a connected spanning subgraph that contains no circuit, that is, a spanning tree. □

Whenever an argument will be used over and over again, it becomes practical to formulate it as a lemma even if the lemma itself is not particularly interesting. The next two lemmas are of this type. The first is a very useful result, one that seems obvious but nevertheless needs to be proved.

Lemma 2.8. *Every forest that has at least one edge has a pendant vertex.*

Proof. Let (V, E) be a forest and let $a = a_0, a_1, \ldots, a_k = b$ be a longest path in (V, E). Since $E \neq \emptyset$, $k \geq 1$ and $a \neq b$. Now b has an edge to a_{k-1}. Suppose b had an edge to some other vertex $c \in V$. If c is not on the path, $a = a_1, a_2, \ldots, a_k = b, c$ is a longer path, contrary to our assumption. On the other hand, if c is on the path, say $c = a_i$, then $c, a_i, \ldots, a_k = b, c$ is a circuit. Since both options are impossible, b is the endpoint of just one edge; that is, b is pendant. □

Lemma 2.9. *Let (V, E) be a connected graph, let a be a pendant vertex and let $e = \{a, b\}$ be the only edge with a as endpoint. Then the graph $(V - a, E - e)$ (obtained by deleting a and e) is connected. In particular, if (V, E) is a tree, then $(V - a, E - e)$ is a tree.*

Proof. Let $c, d \in V - a$. Since (V, E) is connected, there is a path joining c and d in (V, E): $c = c_1, \ldots, c_k = d$. Since, for $i = 2, \ldots, k - 1$, c_i has valence at least 2, none of these vertices can equal a. Hence the entire path lies in $(V - a, E - e)$ and we may conclude that $(V - a, E - e)$ is connected. Finally, if (V, E) contains no circuit, clearly $(V - a, E - e)$ contains no circuit.□

Several of the more useful results about trees can be condensed into the following lemma.

Lemma 2.10. *If a graph (V, E) satisfies any two of the following conditions, then it satisfies all three conditions:*

1. *(V, E) is connected.*
2. *(V, E) contains no circuit.*
3. *$|E| = |V| - 1$.*

Before we prove this lemma let's consider the ways that we might use it. If (V, E) satisfies conditions (1) and (2), it is by definition a tree. So, we conclude that each tree has one more vertex than edge. We can also use this result to conclude that a connected graph with one more vertex than edge is a tree or that a circuit-free graph with one more vertex than edge is also a tree.

Proof. Assume that (V, E) satisfies conditions (1) and (2); that is, it is a tree. We prove that condition (3) holds by induction on $|V|$. Clearly, when $|V| = 1$, $|E| = 0$ and the result holds. Now suppose condition (3) holds for all trees with fewer than n vertices and let (V, E) be a tree with $|V| = n$. By Lemma 2.8, (V, E) has a pendant vertex and by Lemma 2.9 (V', E'), the subgraph obtained by deleting this pendant vertex and its edge, is also a tree. By the induction hypothesis, $|E'| = |V'| - 1$. Substituting $|E| - 1$ for $|E'|$ and $|V| - 1$ for $|V'|$ into this formula gives $|E| = |V| - 1$, as required.

Next assume (V, E) satisfies conditions (2) and (3). By (2), (V, E) is a forest. Let $(V_1, E_1), \ldots, (V_k, E_k)$ denote the trees (components) of this forest. By the first part of this proof, we have $|E_i| = |V_i| - 1$, for $i = 1, \ldots, k$. Summing these equations, we get

$$\begin{aligned} |E| &= |E_1| + \cdots + |E_k| = (|V_1| - 1) + \cdots + (|V_k| - 1) \\ &= |V_1| + \cdots + |V_k| - k = |V| - k. \end{aligned}$$

But $|E| = |V| - 1$! We conclude that $k = 1$ and (V, E) is connected.

Finally, assume that that (V, E) satisfies conditions (1) and (3) and again proceed by induction on $|V|$. Again, when $|V| = 1$, $|E| = 0$ and the result clearly holds. We assume $n > 1$. Using condition (3), we may compute the average valence,

$$\overline{\rho} = \frac{2|E|}{|V|} = \frac{2|V| - 2}{|V|} = 2 - \frac{2}{|V|}.$$

Hence, the average valence is less than 2 and so there must be at least one vertex of valence less than 2. Since (V, E) is connected and has more than one vertex, it has no isolated vertices. We conclude that (V, E) has a pendant vertex. If we delete the pendant vertex to get (V', E'), we can conclude (from Lemma 2.9) that (V', E') is connected and (from a counting argument similar to the one given above) that $|E'| = |V'| - 1$. Thus, by the induction hypothesis, (V', E') contains no circuit. Finally, we note that a pendant vertex can belong to no circuit; thus, if (V', E') is circuit-free, so is (V, E). □

The final result we prove in this section is one we will need later and to state it we need some additional notation. Let (V, E) be a graph. For $U \subseteq V$, let $E(U)$ denote the collection of edges in E that have both endpoints in U.

Lemma 2.11. *A graph (V, E) is circuit-free if and only if $|E(U)| \leq |U| - 1$, for all $\emptyset \subset U \subseteq V$.*

Proof. If (V, E) is circuit-free then so is every subgraph, in particular any subgraph of the form $\big(U, E(U)\big)$. Thus $\big(U, E(U)\big)$ is a forest and, as we have just computed, $|E(U)| = |U| - k \leq |U| - 1$.

Now assume that U is the vertex set of a circuit in (V, E). Since the number of edges in a circuit equals the number of vertices, $|E(U)| \geq |U|$. Hence, if $|E(U)| \leq |U| - 1$ for all $\emptyset \subset U \subseteq V$, (V, E) cannot contain a circuit. □

Listed below are some of the interesting facts about trees and circuits that we will not use and therefore have not included in our development. Their proofs are no more difficult than those given above but could still prove challenging for the interested reader.

Result 2.12. *A graph (V, E) is a tree if and only if every two vertices in (V, E) are joined by a unique path.*

Result 2.13. *A connected graph is a tree if and only if its average valence is less than* 2.

Result 2.14. *A connected graph contains exactly one circuit if and only if its average valence equals* 2.

Result 2.15. *Adding an edge between two vertices of a tree results in a graph containing exactly one circuit.*

2.5 1-Dimensional Rigidity (Part II)

We now have the tools to formulate and prove the main result of 1-dimensional rigidity:

Theorem 2.16. *A* 1*-dimensional framework $(V, E, \mathbf{p})$ is rigid if and only if its structure graph, (V, E), is connected.*

Proof. Let $\mathcal{F} = (V, E, \mathbf{p})$ be a 1-dimensional framework. Assume that (V, E) is disconnected and that (V_1, E_1) is a component of (V, E), Assume that the vertices have been labeled $a_1, \ldots, a_n$ so that $V_1 = \{a_1, \ldots, a_k\}$. Define the motion $\mathbf{P}_i : [0, 1] \to R$ as follows:

- $\mathbf{P}_i(t)$ is the point with coordinate x_i, for $i \leq k$; and
- $\mathbf{P}_i(t)$ is the point with coordinate $x_i + t$, for $i > k$.

One easily checks that conditions (1) and (2) of the definition of a motion are satisfied. If $(a_i, a_j) \in E$, then either i and j are both less than or equal to k or they are both greater than k. Condition (3) holds directly if they are both less than or equal to k. If they are both greater than k,

$$|\mathbf{P}_i(t), \mathbf{P}_j(t)| = |(x_i + t) - (x_j + t)| = |x_i - x_j| = |\mathbf{p}_i, \mathbf{p}_j|;$$

so condition (3) holds in that case too. Finally,

$$|\mathbf{P}_{k+1}(t), \mathbf{P}_1(t)| = |(x_{k+1} + t) - x_1| \neq |x_{k+1} - x_1| = |\mathbf{p}_{k+1}, \mathbf{p}_1|,$$

for $t > 0$. So this is a deformation of $\mathcal{F}$. We have proved that, if (V, E) is not connected, then $(V, E, \mathbf{p})$ is not rigid; or, equivalently, if $(V, E, \mathbf{p})$ is rigid, then (V, E) is connected.

To prove the converse, assume that (V, E) is connected and that $\{\mathbf{P}_i : [0, 1] \to R\}$ is a motion of $\mathcal{F}$. Let $\{a_i, a_j\} \in E$; then,

$$|\mathbf{P}_j(t), \mathbf{P}_i(t)| = |\mathbf{p}_j, \mathbf{p}_i| = |x_j - x_i|,$$

for all $t \in [0, 1]$. This means that $\mathbf{P}_j(t) - \mathbf{P}_i(t)$ can only take on the values $x_j - x_i$ and $x_i - x_j$ (or the single value 0, if $x_i = x_j$). But, $\mathbf{P}_j(t) - \mathbf{P}_i(t)$ is a continuous function and cannot jump between two values. Hence,

$$\mathbf{P}_j(t) - \mathbf{P}_i(t) = x_j - x_i,$$

for all $t \in [0, 1]$. It follows that $\mathbf{P}_j(t) - x_j = \mathbf{P}_i(t) - x_i$, for all $t \in [0, 1]$, whenever $\{a_i, a_j\} \in E$.

Now let a_i and a_j be any two vertices. Since (V, E) is connected, a_i and a_j are joined by a path $a_i = a_{i_0}, a_{i_1}, \ldots, a_{i_k} = a_j$. But then,

$$\mathbf{P}_i(t) - x_i = \mathbf{P}_{i_1}(t) - x_{i_1} = \cdots = \mathbf{P}_j(t) - x_j,$$

for all $t \in [0, 1]$. Thus,

$$|\mathbf{P}_j(t), \mathbf{P}_i(t)| = |\mathbf{P}_j(t) - \mathbf{P}_i(t)| = |x_j - x_i| = |\mathbf{p}_j, \mathbf{p}_i|,$$

for all $t \in [0,1]$, and $\{\mathbf{P}_i : [0,1] \to R\}$ is a rigid motion of $\mathcal{F}$. Hence, when (V, E) is connected, all motions of $\mathcal{F}$ are rigid motions. □

Suppose that a framework $(V, E, \mathbf{p})$ (in any dimension) is rigid:

- If the removal of any segment results in a framework that is not rigid, $(V, E, \mathbf{p})$ is called an *isostatic* framework.

Any rigid framework $(V, E, \mathbf{p})$ that is not isostatic must contain a segment that is unnecessary, that is, a segment whose removal leaves the framework rigid. In general:

- A segment of a framework is said to be an *implied segment* if deleting it leaves a framework with the same collection of motions as before.

Suppose that a framework $(V, E, \mathbf{p})$ (in any dimension) is rigid. If it has an implied segment, that segment may be deleted leaving a rigid framework with one fewer segment. Repeating this reduction as often as necessary, we arrive at an isostatic framework $(V, F, \mathbf{p})$ with $F \subseteq E$. Thus, every rigid framework is obtained by adding superfluous (implied) segments to an isostatic framework. Therefore, it is reasonable to concentrate our investigations on the isostatic frameworks. Returning to 1-dimensional frameworks, two results follow from Lemmas 2.6 and 2.7:

Lemma 2.17. *A 1-dimensional framework is isostatic if and only if its structure graph is a tree.*

Lemma 2.18. *A segment of a 1-dimensional framework is implied if and only if the corresponding edge of its structure graph lies on a circuit.*

We complete this section by proving two characterizations of trees that will generalize to the graphs of isostatic frameworks in dimension two. The first of these characterizations, Theorem 2.19, follows at once from Lemmas 2.10 and 2.11.

Theorem 2.19. *A graph (V, E) is a tree if and only if*

1. $|E(U)| \leq |U| - 1$, *for all* $\emptyset \subset U \subseteq V$, *and*
2. $|E| = |V| - 1$.

This result relates nicely to our degree of freedom approach. The second condition indicates that any framework with (V, E) as structure graph has enough segments to be rigid, while the first condition requires that each segment be properly placed: No segment may be placed between two points of an already rigid piece.

The next result gives a method for constructing all trees. Let the graph (V, E) be given with $a \in V$, $b \notin V$ and $e = \{a, b\}$; then the graph $(V \cup \{b\}, E \cup \{e\})$ is called a 1-*extension* of (V, E).

Theorem 2.20. *A graph (V, E) is a tree if and only if it can be constructed from a single point by a sequence of* 1-*extensions.*

Proof. Assume that (V, E) is a tree, with $a \in V$, $b \notin V$ and $e = \{a, b\}$. We must show that $(V \cup \{b\}, E \cup \{e\})$ is also a tree. Clearly, $(V \cup \{b\}, E \cup \{e\})$ contains no circuit: No circuit can contain b since it has only one neighbor, and any circuit not containing b would have to be a circuit in the tree (V, E). We also have

$$\begin{aligned} |E \cup \{e\}| &= |E| + 1 \\ &= |V| - 1 + 1 \\ &= |V| + 1 - 1 \\ &= |V \cup \{b\}| - 1. \end{aligned}$$

So by Lemma 2.10, $(V \cup \{b\}, E \cup \{e\})$ is a tree.

Since every tree on more than one vertex has a pendant vertex (Lemma 2.8) whose deletion yields a smaller tree (Lemma 2.9), an elementary inductive argument shows that every tree can be constructed by a sequence of 1-extensions. □

One may wish to consider rigid frameworks that have the additional property that they remain rigid even if any single rod fails (is removed). Such frameworks are said to be *birigid*. In view of Lemma 2.6, birigid frameworks may be easily characterized:

Theorem 2.21. *A* 1-*dimensional framework $(V, E, \mathbf{p})$ is birigid if and only if*

1. *(V, E) is connected and*
2. *each edge lies on a circuit.*

Exercise 2.1. *Suppose that the* 1-*dimensional framework $(V, E, \mathbf{p})$ is birigid and bipartite with $V = A \cup B$, where $|A| = n$ and $|B| = m$ as bipartition. How small can E be?*

2.6 The Solution to the Grid Problem

In Chapter 1, we studied the grid bracing problem using our intuitive degrees of freedom model. Now that we have defined the standard model for rigidity, should we try to apply it to grid bracing? As we will demonstrate at the beginning of Chapter 3, this standard model is already complicated to use in the case of the 1×1 grid. It cannot be used to solve the general grid bracing problem.

In 1977, Edwin Bolker and Henry Crapo developed a very special model for the grid bracing problem that capitalizes on the very special structure of these grids ("How to Brace a One-Story Building," *Environment and Planning*, B, 4, 125–52). Their model is far more abstract than the models we have studied so far. Nevertheless, it gives a complete and quite elegant solution to the grid bracing problem.

Let $\mathcal{G}_{nm}$ denote the $n \times m$ grid, that is, the grid with n rows and m columns. Label the rows by $r_1, r_2, \ldots, r_n$ (from top to bottom) and the columns by $c_1, c_2, \ldots, c_m$ (from left to right). We associate with the grid a graph with $n+m$ vertices, indexed by the same symbols, but having no edges. In Figure 2.14, we depict the 3×4 grid and associated graph.

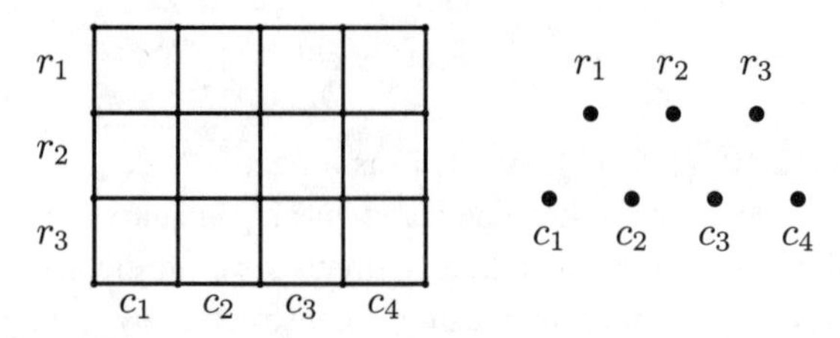

FIGURE 2.14

We can now encode any bracing of the grid as a bipartite graph on this vertex set as follows: If the cell in row r_i and column c_j is braced, the vertices labeled r_i and c_j are joined by an edge. We illustrate this in Figure 2.15.

Despite its simplicity, the bipartite graph that we associate with a braced grid is a complete model of that braced grid, in that it encodes all of the information needed to decide whether the bracing is rigid or not. From the associated bipartite graph, we may reconstruct the grid and identify those cells that are braced. In effect, the associated bipartite graph is a shorthand way to describe a braced grid. But it isn't for its economy that this model is so important. The importance of this mathematical model lies in the following surprising and beautiful result.

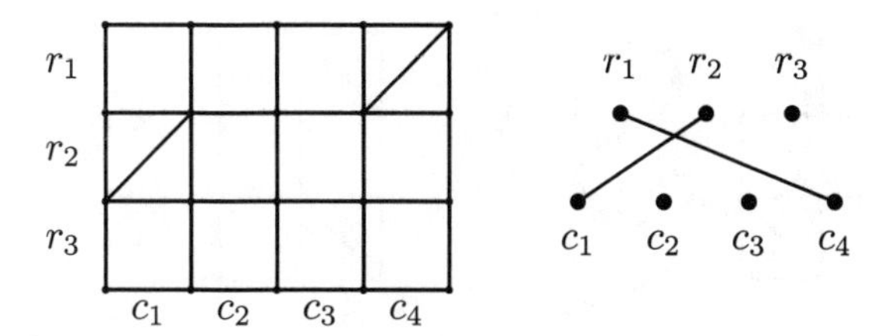

FIGURE 2.15

Theorem 2.22. *A braced grid will be rigid if and only if its associated bipartite graph is connected.*

This is clearly a statement that we should not accept without proof, and we will prove it soon. But for now let's check that it works in some of the simpler cases and then, assuming it is true, see what it can tell us about bracing grids.

We start by checking this statement against what we already know. (Checking that it works in some or even many examples won't prove that it is always true, but it can give us some understanding of how it works, and this understanding will be an aid in understanding just how we might prove it.) The simplest case that we completely understand is the case of the $1 \times n$ grids. Here, as we have noted, every square must be braced. The associated bipartite graph consists of one vertex labeled r_1 in one class and n vertices labeled $c_1, \ldots, c_n$ in the other class. Clearly, such a bipartite graph is connected if and only if every edge between the two classes is included.

Now consider the bipartite graph associated with the $2 \times n$ grid. See Figure 2.16. For this to be a connected graph, it must have at least one edge from each c_i; that is, there must be a brace in each column. However, if there is only one edge from each c_i, the graph is not connected: r_1 and the c_i's connected to it is the vertex set of one component, and r_2 and the c_i's connected to it is the vertex set of the other component. Adding a second edge to one of the c_i's makes the graph connected. Thus, as we have already deduced, a bracing will be rigid if and only if it includes at least one brace in each column and has two braces in at least one column.

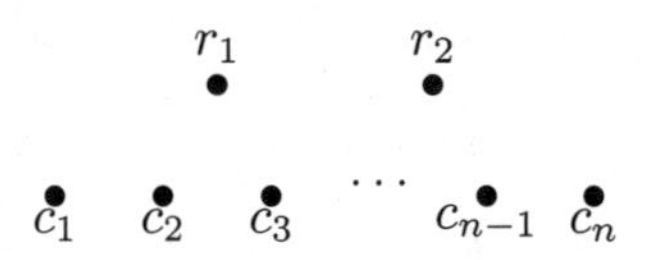

FIGURE 2.16

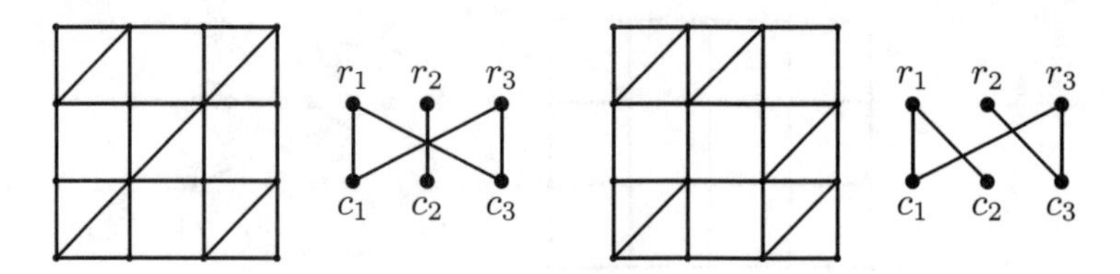

FIGURE 2.17

Recall the two braced 3×3 grids in Figure 1.14; one was rigid and one was not. We reproduce these two braced grids along with their associated graphs in the Figure 2.17.

The associated graph of the right-hand braced grid is connected and that bracing is rigid, as we confirmed in Chapter 1. The associated graph of the left-hand braced grid is not connected: One component is the single edge $\{r_2, c_2\}$ and the other component is the circuit r_1, c_3, r_3, c_1, r_1. A deformation of this nonrigid braced grid was described in the discussion following Figure 1.19. Reviewing the argument showing that this braced grid is not rigid, we see that the $3, 3$ cell must remain square under deformations of the grid even if the brace in that cell is left out. In terms of the associated graph with the edge (r_3, c_3) removed, it seems that the path from r_3 to c_3 has the same effect as the missing edge. With that observation in mind, we now give a formal proof of Theorem 2.22.

Proof. Consider any $m \times n$ grid with the usual labeling of its rows and columns. Referring to an undeformed drawing of it, we will call the vertical rods in the ith row the rods of row i. Similarly, the horizontal rods in column j will be called the rods of column j. The first observation in this proof is the fact that, in any deformation, each cell is a parallelogram. It follows from this observation that, no matter how the grid is deformed, the rods of row i remain parallel to one another. Similarly, the rods of column j remain parallel to one another. And this is true for each row index i and each column index j. This observation is illustrated in Figure 2.18.

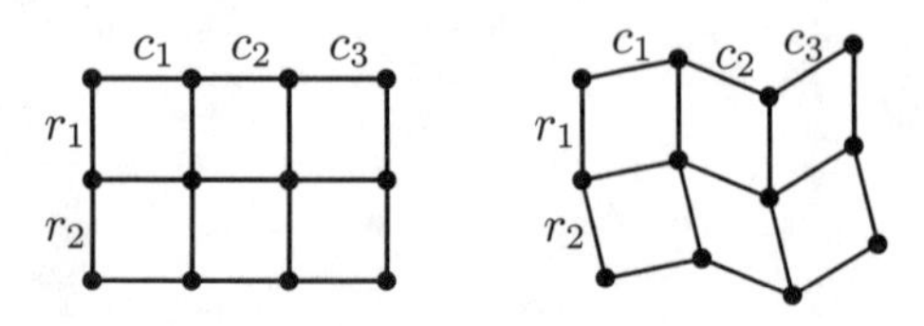

FIGURE 2.18

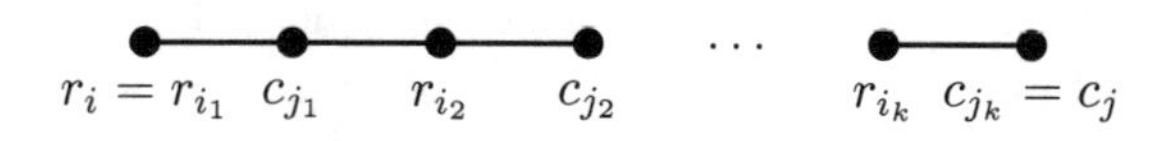

FIGURE 2.19

Suppose that there is a brace in the i, j cell of the grid, that is, an edge from r_i to c_j in the associated graph. Since, in any deformation of the braced grid, the rods of row i that bound the braced cell are perpendicular to the rods of column j that bound the braced cell, we conclude that all the rods of row i are perpendicular to all the rods of column j. Suppose that braces have been added so that there is a path in the associated bipartite graph from r_i to c_j. See Figure 2.19.

We argue as follows: Since the rods of row r_{i_1} and the rods of row r_{i_2} are perpendicular to the rods of column c_{i_1}, the rods of row r_{i_1} are parallel to the rods of row r_{i_2}. Similarly, the rods of column c_{i_1} are parallel to the rods of column c_{i_2}. Inductively,

- all the rods in all the rows of the path are parallel to one another;
- all the rods in all the columns of the path are parallel to one another; and
- all the rods in all the rows of the path are perpendicular to all the rods in all the columns of the path.

In particular, the row rods bounding the i, jth cell are perpendicular to the column rods bounding the i, jth cell.

In short, if there is a path from r_i to c_j in the associated graph, the cell in the ith row and jth column must remain square under all motions of the braced grid. Thus, if the associated graph is connected, then there is a path from each r_i to each c_j and each cell must remain square under all motions; that is, the braced grid has no deformations.

To prove the converse, assume that $\mathcal{G}$ is an $m \times n$ braced grid whose associated graph is not connected. We must show that $\mathcal{G}$ admits a deformation. Let A denote the set of vertices of the component of the associated bipartite graph that contains r_1, and let B be the set of the remaining vertices. Hence, if e is an edge either both of its endpoints are in A or both are in B. Equivalently, if the i, jth cell is braced, then either r_i and c_j are both in A or they are both in B.

Recall that, in Section 1.5, we observed that the edges along the top and left side of the unbraced grid may be independently reoriented. We use this observation to construct a deformation of this braced grid.

Let α denote the measure of some small angle and adjust the rods along the left side and top of the unbraced grid as follows:

- If $r_i \in A$, the corresponding rod is vertical;
- If $c_j \in A$, the corresponding rod is horizontal.
- If $r_i \in B$, the corresponding rod makes a counterclockwise angle of measure α with the vertical.
- If $c_j \in B$, the corresponding rod makes a counterclockwise angle of measure α with the horizontal.

We illustrate this with the 3×5 grid in Figure 2.20, where

$$A = \{r_1, r_3, c_2, c_3, c_5\},$$

$$B = \{r_2, c_1, c_4\}.$$

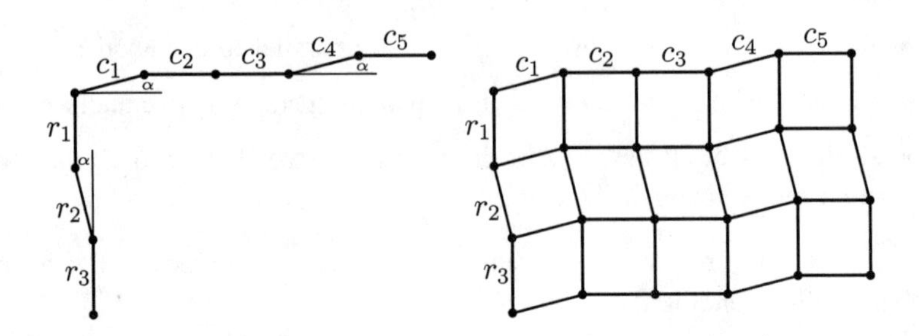

FIGURE 2.20

One easily sees that cells of the deformed grid whose row and column vertices are in the same set are square! Thus all of the cells that could be braced are square. Hence, the undeformed braced grid can be deformed by increasing α from 0 to some positive value. □

Theorem 2.22 has several consequences that follow from it and the results we have already proved about connectivity and trees. The first of these results was conjectured earlier (Conjecture 1.1).

Corollary 2.23. *The $n \times m$-grid needs at least $n+m-1$ braces to be rigid and can be made rigid with $n+m-1$ properly placed braces.*

Corollary 2.24. *The $n \times m$-grid with $n+m-1$ braces will be rigid if and only if the associated bipartite graph is a tree.*

Corollary 2.25. *A rigid bracing of the $n \times m$-grid will remain rigid upon the deletion of any one brace if and only if the associated bipartite graph is connected and has the property that every edge lies on a circuit.*

The reader interested in developing a good understanding of the grid problem should prove each of these corollaries. Also, for the interested reader, we leave unresolved several natural questions about the structure of redundant bracings. For example:

Exercise 2.1. *What is the minimum number of braces needed for a rigid bracing of the $n \times m$ grid that remains rigid upon the deletion of any one brace?*

The Bolker–Crapo mathematical model for the grid bracing problem actually reduces the 2-dimensional grid rigidity problem to a 1-dimensional rigidity (graph connectivity) problem. Unfortunately, no such reduction is known to exist for the general 2-dimensional rigidity problem.

Exercise 2.2. *Suppose that a set of five of the nine cells of $\mathcal{G}_{33}$ is chosen at random and braced. What is the probability that the resulting braced grid will be rigid?*

2.7 Planar Graphs

This section is devoted to developing the theory of graphs that can be drawn in the plane without their edges crossing and to proving one of the key results of this theory. However, we will not use the material in this section until we get toward the end of Chapter 4. Hence, the reader could put off reading this section until its application arises.

A graph (V, E) is said to be *planar* if it can be drawn in the plane without its edges crossing. Of the complete graphs, K_1, K_2, K_3 and K_4 are planar. Among the complete bipartite graphs, $K_{1,n}$ and $K_{2,n}$ are planar, for all n. Planar drawings of K_4 and $K_{2,n}$ are given in Figure 2.21.

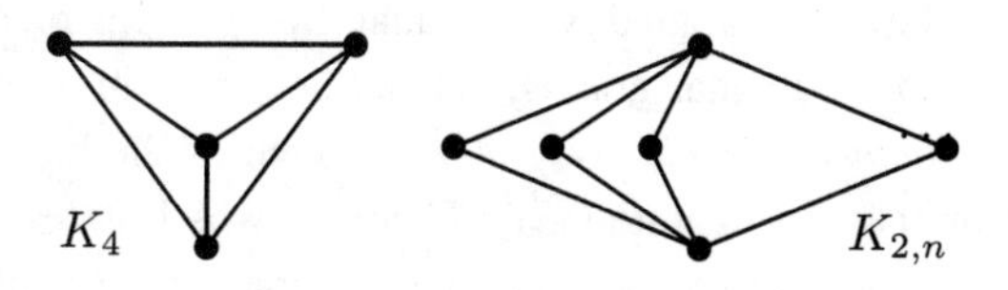

FIGURE 2.21

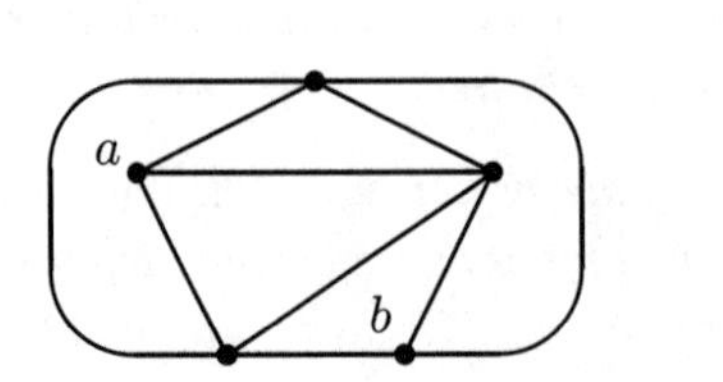

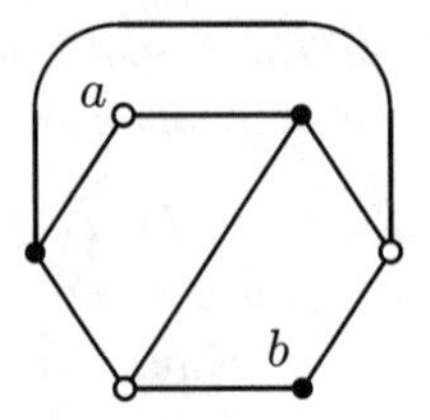

FIGURE 2.22

The simplest nonplanar graphs are K_5 and $K_{3,3}$. In Figure 2.22, we draw these graphs in the plane with one edge deleted. We use these drawings to demonstrate that K_5 and $K_{3,3}$ are not planar. The arguments are based on the following simple observation: If all of the vertices of a graph lie on a circuit, then all planar embeddings can be obtained by first drawing the circuit and then adding the remaining edges.

- K_5: In this case once the pentagon has been drawn, it is easy to see that at most two of the remaining five edges can be drawn inside the pentagon without crossing and at most two can be drawn outside the pentagon without crossing. This leaves one edge ($\{a, b\}$ in this drawing) that cannot be included without a crossing.

- $K_{3,3}$: Here we have drawn the hexagon with open circles representing one class of the bipartition of the vertex set. We see that the remaining three edges are the diagonals. But only one diagonal can be drawn inside the hexagon without a crossing and only one diagonal can be drawn outside without a crossing. This leaves one edge ($\{a, b\}$ in this drawing) that cannot be included without a crossing.

One can make any number of nonplanar graphs from these two graphs. Take either one and subdivide the edges (replace each edge by a path). Clearly, any such graph cannot be drawn in the plane without giving a planar drawing for the underlying K_5 or $K_{3,3}$. Now add as many additional vertices and edges as you wish to this graph. If the resulting graph had a planar drawing, throwing away the added vertices and edges would leave a planar drawing of the graph that we have just argued is nonplanar. So this construction yields an infinite collection of nonplanar graphs that, we say, contain a *homeomorphic* copy of K_5 or a *homeomorphic* copy of $K_{3,3}$. One of the best known early theorems in graph theory is Kuratowski's Theorem, which states the surprising (and not so easy to verify) fact that *every* nonplanar graph contains either a homeomorphic copy of K_5 or a homeomorphic copy of $K_{3,3}$.

Once a graph has been drawn in the plane it acquires additional features called *faces*: the regions of the plane bounded by the drawing of the graph in the plane. If the graph is connected, all faces (except the outside face) are simply connected (distorted disks).

Actually, we may avoid the special treatment of the outside face if the drawing is made on the sphere instead of the plane. In a spherical drawing, the "outside" face closes over the back side of the sphere to also form a disk-like face. Since it is very convenient to avoid dealing with an exceptional outside face, we adopt the common practice of always assuming that our drawings are on the sphere while continuing to use the terminology of planar embeddings.

A graph may be drawn in the plane in many different ways—but some ways are "more different" than others. First of all, we could simply distort the picture without changing the relationships of the various parts. For example, in Figure 2.23, we distort the planar graph on the left by pushing vertices a and d in and down and straightening the arcs representing $\{c, e\}$ and $\{c, b\}$, to eventually get the straight-line drawing on the right.

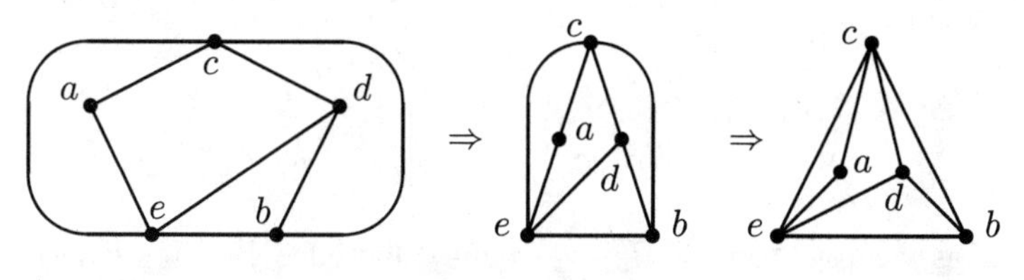

FIGURE 2.23

To distinguish the combinatorial properties of a drawing of a graph in the plane from the geometric properties, we identify each face with its *boundary list*—that is, the list of vertices, in order, around the face—instead of with the actual disk-like region. Combinatorially speaking, the three drawings in Figure 2.23 are the same since they have the same boundary lists:

$$f_1:\ a, c, e, a;\quad f_2:\ a, e, d, c, a;\quad f_3:\ d, e, b, d;$$
$$f_4:\ c, d, b, c;\quad f_5: c, e, b, c.$$

As in the listing of a circuit, the first and last vertex in the lists are the same. If no other vertices are repeated, as in the boundary lists for faces of this example, the boundary list is a circuit.

However, this need not always be the case. Consider the graph in Figure 2.24. The boundary lists for f_1 (a_1, a_2, a_3, a_1) and f_2 (a_4, a_5, a_6, a_4)

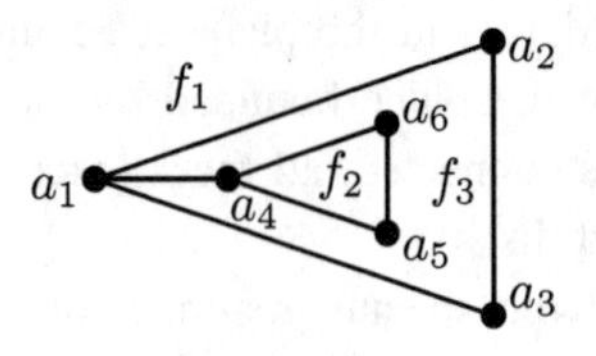

FIGURE 2.24

are circuits, but the boundary list for f_3 (a_1, a_4, a_6, a_5, a_4, a_1, a_3, a_2, a_1) is not.

Even when a boundary list is not a circuit we call the number of steps needed to get around the face (the number of vertices in the list minus one) the *valence* of the face. We use the notation $\rho(f)$ to denote the valence of the face f.

The valence of a face can also be thought of as the number of edges surrounding the face. For example, in the planar graph of Figure 2.24, the faces f_1 and f_2 are both surrounded by three edges while f_3 is surrounded by eight edges:

$$\begin{array}{cccc} \{a_1, a_4\}, & \{a_4, a_6\}, & \{a_6, a_5\}, & \{a_5, a_4\}, \\ \{a_4, a_1\}, & \{a_1, a_3\}, & \{a_3, a_2\}, & \{a_2, a_1\}. \end{array}$$

Notice, as we count around f_3, we counted the edge $\{a_1, a_4\}$ twice—once from each side. If we add up all of the face valences $(3 + 3 + 8 = 14)$, we have counted each edge exactly twice—once from each side. Thus, we have a face-valence version of Lemma 2.1.

Lemma 2.26. *Let any planar embedding of the graph (V, E) be given and let F denote the set of all faces of that embedding. Then*

$$\sum_{f \in F} \rho(f) = 2|E|;$$

and the average face-valence of the embedding is

$$\hat{\rho} = \frac{2|E|}{|F|}.$$

In Figure 2.25, we give three drawings of the same graph. They look quite different. However, when we write out their boundary lists, we see that the

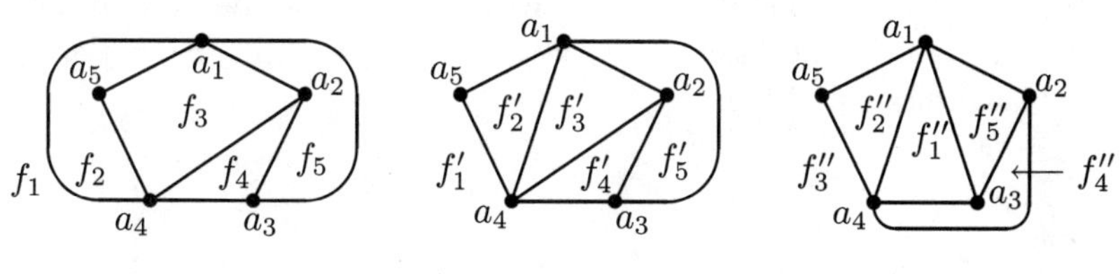

FIGURE 2.25

first and the last are the same, combinatorially speaking, while the central drawing has different boundary lists.

- First drawing:
  $f_1: a_1, a_3, a_4, a_1;\quad f_2: a_5, a_1, a_4, a_5;\quad f_3: a_1, a_5, a_4, a_2, a_1$
  $f_4: a_2, a_4, a_3, a_2;\quad f_5: a_1, a_2, a_3, a_1.$

- Second drawing:
  $f'_1: a_1, a_3, a_4, a_5, a_1;\quad f'_2: a_5, a_1, a_4, a_5;\quad f'_3: a_1, a_4, a_2, a_1;$
  $f'_4: a_2, a_4, a_3, a_2;\quad f'_5: a_1, a_2, a_3, a_1.$

- Third drawing:
  $f''_1: a_1, a_3, a_4, a_1;\quad f''_2: a_5, a_1, a_4, a_5;\quad f''_3: a_1, a_5, a_4, a_2, a_1;$
  $f''_4: a_2, a_3, a_4, a_2;\quad f''_5: a_1, a_2, a_3, a_1.$

The first and third drawings have the same set of boundary lists, and the faces have been labeled to demonstrate this. Since they have the same set of boundary lists, we say they are the same *planar embedding* of the graph. The second drawing is a different planar embedding of the graph. This can be seen by observing that the boundary list of f'_1, a_1, a_3, a_4, a_5, a_1 is not a boundary list for either of the other drawings.

The term *planar graph* refers to a graph that can be drawn in the plane without crossings; the terms *plane graph* and *map* are used interchangeably to denote a planar graph along with a specific drawing in the plane. We lump together combinatorially equivalent maps (that is, those with the same boundary lists) and denote them by (V, E, F), where F is the collection of faces. So, while only one planar graph is exhibited in Figure 2.25, two distinct plane graphs or maps are represented. Note that the numbers of faces in all three maps are the same. This is no accident—as the next famous theorem, due to Euler, shows.

Theorem 2.27. *Let (V,E,F) be a connected plane graph. Then we have that* $|V| - |E| + |F| = 2.$

Proof. We observe that the result holds for trees. Every planar drawing of a tree has exactly one face; so $|F| = 1$. By Lemma 2.10, $|E| = |V| - 1$. Combining these, we have:

$$|V| - |E| + |F| = |V| - (|V| - 1) + 1 = 2.$$

We proceed by induction on $|E|$. We assume that, if (V', E', F') is a connected plane graph and $|E'| < n$, then $|V'| - |E'| + |F'| = 2$. We also assume that (V, E, F) is a connected planar graph with $|E| = n$. If (V, E) is a tree, we have already shown that the formula holds; so assume (V, E) is not a tree.

Since (V, E) is connected but not a tree, it must contain a circuit. Let $e \in E$ lie on a circuit and consider $(V, E - e)$. By Lemma 2.6, $(V, E - e)$ is still connected. It is still drawn in the plane, and we denote its faces by F'. Since e lies on a circuit, the faces in F on either side of e are different—one lies inside the drawing of the circuit and one outside. Removing e combines these faces into one and leaves all other faces intact. Hence, $|F| = |F'| + 1$. Furthermore, the planar graph (V', E', F'), where $V' = V$ and $E' = E - e$, has fewer than n edges. Combining these facts and our induction hypothesis, we have that

$$\begin{aligned} |V| - |E| + |F| &= |V'| - (|E'| + 1) + (|F'| + 1) \\ &= |V| - |E'| + |F'| = 2. \end{aligned}$$

This completes the proof. □

Euler's formula, $|V| - |E| + |F| = 2$, is one of the most powerful and useful formulas in all of graph theory. It plays a critical role in the proof of the famous Four-Color Theorem, which states that the faces of any map may be colored with just four colors so that no two faces sharing a common boundary edge are assigned the same color. It can also be used to give a purely combinatorial proof that there are exactly five platonic solids (see Exercise 2.1). We will use Euler's formula later in our consideration of geodesic domes. One interesting observation is that vertices and faces appear in symmetric roles in this formula. The symmetry also appears when comparing Lemmas 2.26 and 2.1. There is a very beautiful explanation for this symmetry.

Consider the map, (V, E, F), pictured on the left in Figure 2.26. Construct a new map as follows: Place a new vertex in the interior of each face; if two faces are separated by an edge, draw in a new edge joining the new vertices assigned to these faces. This construction has been carried out on the right in

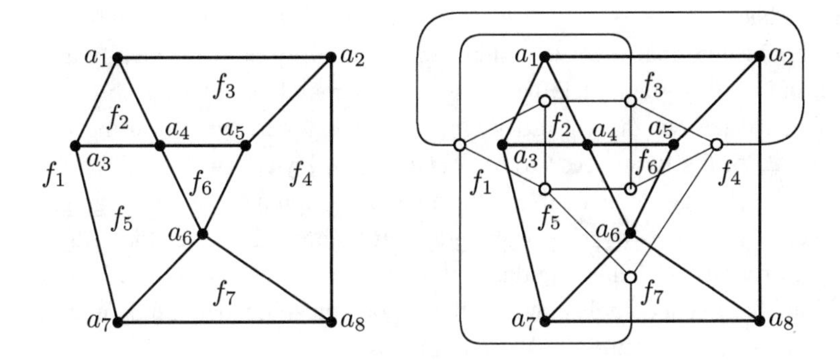

FIGURE 2.26

Figure 2.26. This new map is called the *dual map* to (V, E, F). The vertices of this new map correspond to the faces of the original map, as our labels indicate.

The edges of the two maps have a natural one-to-one correspondence between them: Each edge of the new map corresponds to the edge of the old map that it crosses. For example, consider the edge $e = \{a_6, a_8\}$ of (V, E, F). It separates two faces, f_4 and f_7, so we identify the edge $\{f_4, f_7\}$ of the dual with the edge $\{a_6, a_8\}$ of the original. Now observe that each vertex of (V, E, F) lies in the center of a face of the dual map. Hence the faces of the dual map may be identified with the vertex set V. So the roles of vertices and faces are completely reversed in the dual. Using these identifications, we denote the dual map by (F, E, V). We have drawn (F, E, V) in Figure 2.27.

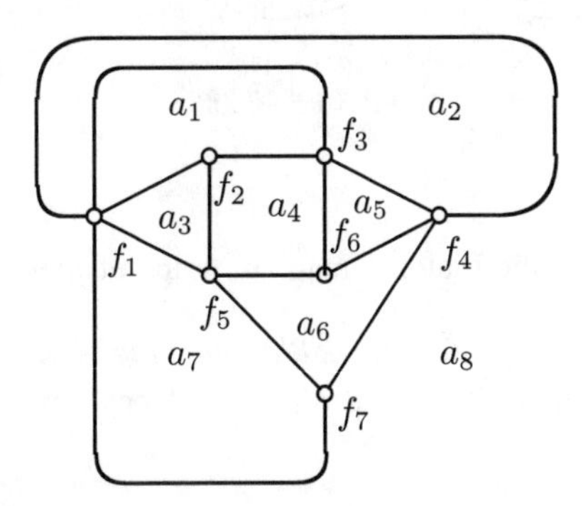

FIGURE 2.27

It should be clear that, if we now construct the dual to (F, E, V), we will get (V, E, F) back.

We should note that there is a slight incompatibility between duality and our definition of a graph. The perceptive reader may have already observed that, if you carry out the construction of the dual for the graph on the left in Figure 2.25, the two edges $\{a_1, a_5\}$ and $\{a_4, a_5\}$ yield two copies of the same edge, $\{f_2, f_3\}$, in the dual. So, the dual of a graph may be a multigraph. Even worse, the edge $\{a_1, a_4\}$ of the graph in Figure 2.24, gives the "edge" $\{f_3, f_3\}$, or a "loop." Thus, the duals of some graphs are not graphs, by our definition. To have a complete duality theory, one must broaden the definition of "graph" to include multiple edges and loops.

The *Platonic solids* are polyhedra all of whose faces are congruent regular polygons, with the same number of polygons coming together at each vertex. That there are just five of them up to similarity (the tetrahedron, the octahedron, the cube, the icosahedron and the dodecahedron) was proved by Euclid. The vertices, edges and faces of a Platonic solid may be interpreted as a map (V, E, F), and these maps are called the Platonic maps. The maps of the cube, octahedron and tetrahedron are drawn in Figure 2.28 and the maps of the icosahedron and dodecahedron are drawn in Figure 2.29.

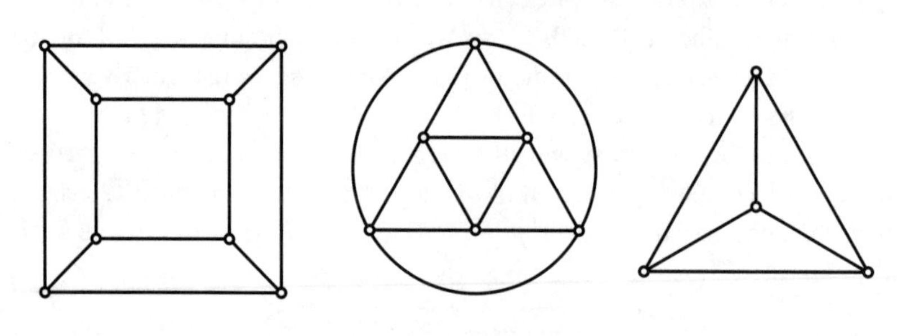

FIGURE 2.28

We formally define the Platonic maps in terms of their regularity:

- A map (V, E, F) is said to be a *Platonic map* if all of its vertices have the same valence, $\rho \geq 3$, and all of its faces have the same valence, $\eta \geq 3$.

Exercise 2.1. *Prove that there are just five Platonic maps.*

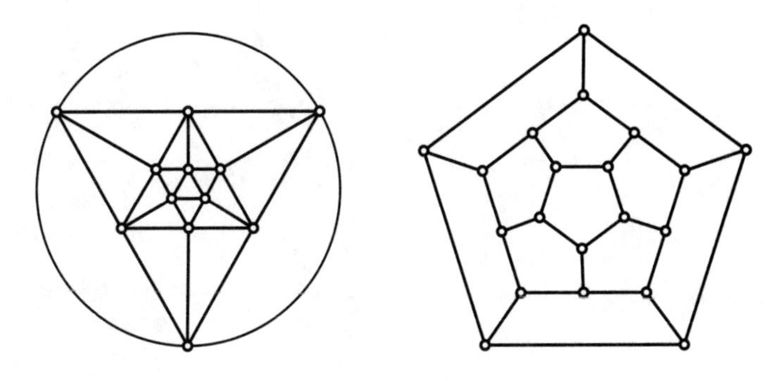

FIGURE 2.29

1. *First use Lemmas 2.1 and 2.26 and Euler's formula to prove*

$$\frac{1}{\rho} + \frac{1}{\eta} = \frac{1}{2} + \frac{1}{|E|}.$$

2. *Next, verify that, under the assumption that ρ and η are integers greater than 2, there are just five solutions of this equation resulting in the following table:*

	ρ	η	$\|E\|$	$\|V\|$	$\|F\|$
tetrahedron	3	3	6	4	4
octahedron	4	3	12	6	8
cube	3	4	12	8	6
icosahedron	5	3	30	12	20
dodecahedron	3	5	30	20	12

3. *Note that the names assigned to the numerical solutions are as yet unwarranted. To complete the proof, you must show that each numerical solution leads to just one planar map. This is most easily done by showing that the drawing of a map corresponding to a numerical solution is completely forced.*

CHAPTER 3

Rigidity Theory

3.1 Rigidity Comes in Two Flavors

For dimensions greater than one, rigidity, as we have defined it, involves a system of quadratic equations. We illustrate with a simple example in two dimensions. Consider the planar[1] framework $(V, E, \mathbf{p})$, where V is the set of four vertices $\{a_1, a_2, a_3, a_4\}$; E denotes the five edges $\{a_1, a_2\}$, $\{a_2, a_3\}$, $\{a_3, a_4\}$, $\{a_1, a_4\}$ and $\{a_1, a_3\}$; and $\mathbf{p}_1 = (0, 0)$, $\mathbf{p}_2 = (3, 0)$, $\mathbf{p}_3 = (3, 2)$ and $\mathbf{p}_4 = (0, 2)$. We have pictured this framework to the left in Figure 3.1.

We wish to prove this framework to be rigid. Since it consists of two triangles, we could use simple geometric arguments to do this. However, many rigid frameworks are not made up of triangles and such geometric arguments do not apply. Furthermore, our purpose here is to demonstrate the algebraic approach to verifying rigidity, and this simple example is just a vehicle for doing so.

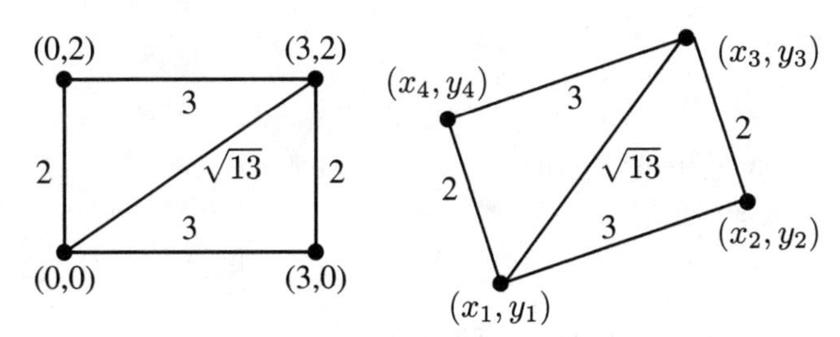

FIGURE 3.1

[1] The terms *planar graph* and *planar framework* can be misleading. The "planar" in planar framework refers to the fact that the points and segments lie in the plane and the framework's motions must take place in the plane, but it does not preclude two of its segments crossing one another. Hence the structure graph of a planar framework need not be a planar graph.

Let $\{\mathbf{P}_i : [0,1] \to R^2 : i = 1, \ldots, 4\}$ be a motion of this framework. On the right side of Figure 3.1, we have pictured the framework at some time $\bar{t} \in [0,1]$ where the coordinates of the point $\mathbf{P}_i(\bar{t})$ are denoted by (x_i, y_i), for $i = 1, \ldots, 4$. We wish to show that the distance between the points (x_2, y_2) and (x_4, y_4) will remain $\sqrt{13}$, even though there is no segment joining them.

In algebraic terms, the coordinates of the points must satisfy the following system of quadratic equations:

$$(x_1 - x_2)^2 + (y_1 - y_2)^2 = 9;$$

$$(x_2 - x_3)^2 + (y_2 - y_3)^2 = 4;$$

$$(x_3 - x_4)^2 + (y_3 - y_4)^2 = 9;$$

$$(x_4 - x_1)^2 + (y_4 - y_1)^2 = 4;$$

$$(x_1 - x_3)^2 + (y_1 - y_3)^2 = 13.$$

And the equation associated with the missing segment,

$$(x_4 - x_2)^2 + (y_4 - y_2)^2 = 13,$$

should be an algebraic consequence of the previous five equations. That this is so is not at all obvious! In fact, proving this is rather complicated even though it involves only standard algebraic operations.

The first simplification is the elimination of the variables x_1 and y_1 by the introduction of a new set of variables. Let

$$\hat{x}_i = x_i - x_1, \quad \text{and} \quad \hat{y}_i = y_i - y_1.$$

Rewriting our system in terms of these variables, we get

$$(\hat{x}_2)^2 + (\hat{y}_2)^2 = 9;$$

$$(\hat{x}_2 - \hat{x}_3)^2 + (\hat{y}_2 - \hat{y}_3)^2 = 4;$$

$$(\hat{x}_3 - \hat{x}_4)^2 + (\hat{y}_3 - \hat{y}_4)^2 = 9;$$

$$(\hat{x}_4)^2 + (\hat{y}_4)^2 = 4;$$

$$(\hat{x}_3)^2 + (\hat{y}_3)^2 = 13.$$

To get the second equation, rewrite $(x_2 - x_3)$ as

$$[(x_2 - x_1) - (x_3 - x_1)] = (\hat{x}_2 - \hat{x}_3).$$

The same trick gives the third equation and

$$(x_4 - x_2)^2 + (y_4 - y_2)^2 = (\hat{x}_4 - \hat{x}_2)^2 + (\hat{y}_4 - \hat{y}_2)^2.$$

So we must now show that $(\hat{x}_4 - \hat{x}_2)^2 + (\hat{y}_4 - \hat{y}_2)^2 = 13$ follows from this new system of quadratic equations.

Geometrically speaking, this change of variables puts $\mathbf{P}_1(\bar{t})$ at the origin of the new coordinate system; the next change rotates the axes so that $\mathbf{P}_2(\bar{t})$ is on the positive $\hat{x}$-axis. Denoting our third set of variables by $\tilde{x}_i$ and $\tilde{y}_i$, let

$$\tilde{x}_i = \left(\frac{\hat{x}_2}{3}\right)\hat{x}_i + \left(\frac{\hat{y}_2}{3}\right)\hat{y}_i \qquad \text{and} \qquad \tilde{y}_i = \left(\frac{-\hat{y}_2}{3}\right)\hat{x}_i + \left(\frac{\hat{x}_2}{3}\right)\hat{y}_i.$$

Then expanding $(\tilde{x}_i - \tilde{x}_j)^2 + (\tilde{y}_i - \tilde{y}_j)^2$, we get

$$\begin{aligned}
&(\tilde{x}_i - \tilde{x}_j)^2 + (\tilde{y}_i - \tilde{y}_j)^2 \\
&\quad = \left[\frac{\hat{x}_2}{3}(\hat{x}_i - \hat{x}_j) + \frac{\hat{y}_2}{3}(\hat{y}_i - \hat{y}_j)\right]^2 + \left[\frac{-\hat{y}_2}{3}(\hat{x}_i - \hat{x}_j) + \frac{\hat{x}_2}{3}(\hat{y}_i - \hat{y}_j)\right]^2 \\
&\quad = \frac{\hat{x}_2^2}{9}(\hat{x}_i - \hat{x}_j)^2 + \frac{\hat{x}_2\hat{y}_2}{9}(\hat{x}_i - \hat{x}_j)(\hat{y}_i - \hat{y}_j) + \frac{\hat{y}_2^2}{9}(\hat{y}_i - \hat{y}_j)^2 \\
&\qquad + \frac{\hat{y}_2^2}{9}(\hat{x}_i - \hat{x}_j)^2 - \frac{\hat{x}_2\hat{y}_2}{9}(\hat{x}_i - \hat{x}_j)(\hat{y}_i - \hat{y}_j) + \frac{\hat{x}_2^2}{9}(\hat{y}_i - \hat{y}_j)^2 \\
&\quad = (\hat{x}_i - \hat{x}_j)^2 + (\hat{y}_i - \hat{y}_j)^2.
\end{aligned}$$

A similar but simpler expansion gives

$$\tilde{x}_i^2 + \tilde{y}_i^2 = \hat{x}_i^2 + \hat{y}_i^2.$$

We also compute

$$\tilde{x}_2 = \frac{\hat{x}_2^2 + \hat{y}_2^2}{3} = 3$$

and

$$\tilde{y}_2 = \frac{-\hat{y}_2\hat{x}_2 + \hat{x}_2\hat{y}_2}{3} = 0.$$

Thus our system of equations has become

$$(\tilde{x}_3 - 3)^2 + (\tilde{y}_3)^2 = 4;$$
$$(\tilde{x}_3 - \tilde{x}_4)^2 + (\tilde{y}_3 - \tilde{y}_4)^2 = 9;$$
$$(\tilde{x}_4)^2 + (\tilde{y}_4)^2 = 4;$$
$$(\tilde{x}_3)^2 + (\tilde{y}_3)^2 = 13.$$

And the equation associated with the missing segment has become

$$\begin{aligned}(x_4 - x_2)^2 + (y_4 - y_2)^2 &= (\hat{x}_4 - \hat{x}_2)^2 + (\hat{y}_4 - \hat{y}_2)^2 \\ &= (\tilde{x}_4 - 3)^2 + (\tilde{y}_4)^2 \\ &= 13.\end{aligned}$$

Solving the first and fourth equations in our new system, we get

$$\tilde{x}_3 = 3 \quad \text{and} \quad \tilde{y}_3 = \pm 2.$$

Substituting these values into the second equation yields the systems

$$(3 - \tilde{x}_4)^2 + (2 - \tilde{y}_4)^2 = 9,$$
$$(\tilde{x}_4)^2 + (\tilde{y}_4)^2 = 4$$

and

$$(3 - \tilde{x}_4)^2 + (-2 - \tilde{y}_4)^2 = 9,$$
$$(\tilde{x}_4)^2 + (\tilde{y}_4)^2 = 4.$$

Solving these systems gives four possible solutions:

$$\begin{aligned}&\tilde{x}_3 = 3, \quad \tilde{y}_3 = 2, \quad \tilde{x}_4 = 0, \quad \tilde{y}_4 = 2;\\ &\tilde{x}_3 = 3, \quad \tilde{y}_3 = 2, \quad \tilde{x}_4 = \tfrac{24}{13}, \quad \tilde{y}_4 = -\tfrac{10}{13};\\ &\tilde{x}_3 = 3, \quad \tilde{y}_3 = -2, \quad \tilde{x}_4 = 0, \quad \tilde{y}_4 = -2;\\ &\tilde{x}_3 = 3, \quad \tilde{y}_3 = -2, \quad \tilde{x}_4 = \tfrac{24}{13}, \quad \tilde{y}_4 = \tfrac{10}{13}.\end{aligned}$$

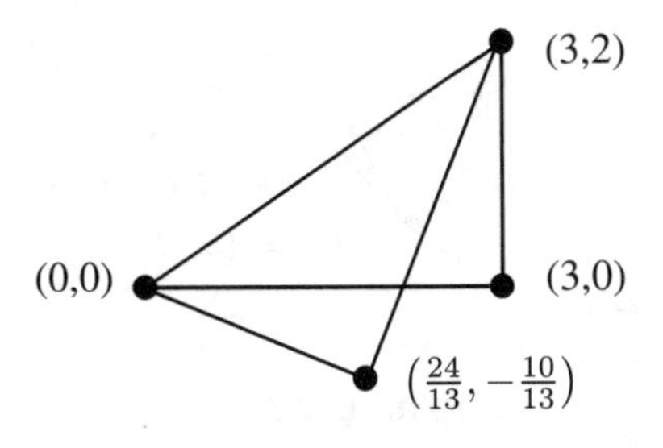

FIGURE 3.2

These solutions yield two distinct values for the square of the distance between $\mathbf{P}_2(\bar{t})$ and $\mathbf{P}_4(\bar{t})$:

$$\begin{aligned}(x_4 - x_2)^2 + (y_4 - y_2)^2 &= (\hat{x}_4 - \hat{x}_2)^2 + (\hat{y}_4 - \hat{y}_2)^2 \\ &= (\tilde{x}_4 - 3)^2 + (\tilde{y}_4)^2 \\ &= 13 \text{ or } \frac{25}{13}.\end{aligned}$$

That both of these values are compatible with the distances fixed by the five segments is clear from Figure 3.2.

We still have not shown that the distance between $\mathbf{P}_2(\bar{t})$ and $\mathbf{P}_4(\bar{t})$ is $\sqrt{13}$ for all t. We complete the argument by noting the following:

- The distance between $\mathbf{P}_2(0) = \mathbf{p}_2$ and $\mathbf{P}_4(0) = \mathbf{p}_4$ is $\sqrt{13}$.
- The distance between $\mathbf{P}_2(t)$ and $\mathbf{P}_4(t)$ is a continuous function of t.
- Thus, the distance between $\mathbf{P}_2(t)$ and $\mathbf{P}_4(t)$ can never jump from $\sqrt{13}$ to $5/\sqrt{13}$.

In geometric terms, no continuous motion *in the plane* can fold the rectangle over its diagonal.

The fact that even such small systems of quadratic equations are so complicated has been part of the motivation for the development of another approach to rigidity called *infinitesimal rigidity*.

Up to this time we have not used the full power of condition (2) of the definition of a motion of a framework. We have used only that the motion of the image of each vertex is continuous. The differentiability requirement implies that, at each position along the trajectory of a point, the velocity vector is well defined; see Figure 3.3. For a vertex $a_i \in V$, the trajectory of

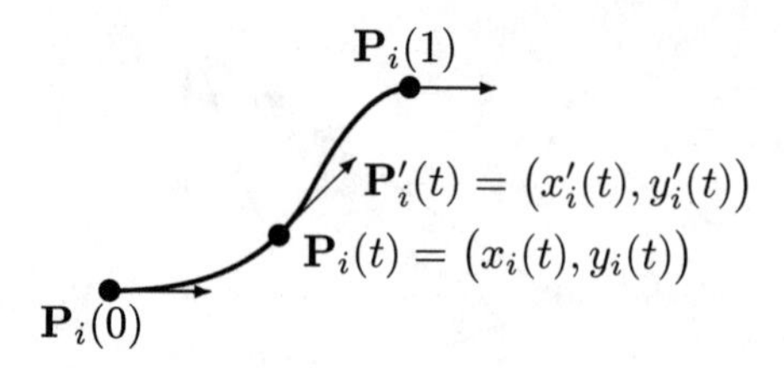

FIGURE 3.3

the corresponding point of the framework is given by $\mathbf{P}_i(t)$. If we let $x_i(t)$ and $y_i(t)$ denote the coordinates of this point at time t, the velocity vector, which we denote by $\mathbf{P}'_i(t)$, is given by

$$(x'(t), y'(t)) = \left(\frac{d}{dt}\,x(t),\ \frac{d}{dt}\,y(t)\right).$$

Let us consider a single segment as it moves in the plane. Let $(V, E, \mathbf{p})$ be a framework consisting of two vertices, $V = \{a_1, a_2\}$, and a single edge $e = \{a_1, a_2\}$. Let $\mathbf{P}_1, \mathbf{P}_2$ be a rigid motion of the segment and, for $i = 1, 2$, let $\mathbf{P}_i(t) = (x_i(t), y_i(t))$. Let $(\overline{x}_1, \overline{y}_1)$ and $(\overline{x}_2, \overline{y}_2)$ be the coordinates of the positions of the endpoints of the segment at some fixed time $\overline{t}$, and let (u_1, v_1) denote the coordinates of the velocity vector of the endpoint $(\overline{x}_1, \overline{y}_1)$ at time $\overline{t}$ while (u_2, v_2) denotes the coordinates of the velocity vector of the endpoint $(\overline{x}_2, \overline{y}_2)$. In short,

$$(\overline{x}_i, \overline{y}_i) = (x_i(\overline{t}), y_i(\overline{t})) \quad \text{and} \quad (u_i, v_i) = (x'_i(\overline{t}), y'_i(\overline{t})).$$

In Figure 3.4, we illustrate several possible motions for this segment.

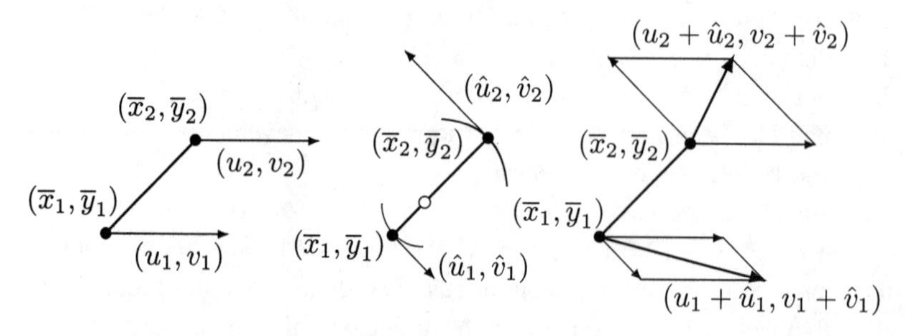

FIGURE 3.4

In the first picture, the motion that has its velocity vectors depicted is a translation. Here the velocity vectors are the same at each endpoint, and the trajectory of each endpoint is a straight line in the direction of the velocity vector. The segment is rotating about the point indicated by the open circle in the second picture, and the trajectories of the endpoints are circles. Both motions are combined in the third picture; think of the segment as tumbling.

What distinguishes the velocity vectors of a rigid motion of a segment from an arbitrary assignment of vectors to the endpoints? The answer lies in the fact that a rigid motion of the segment neither stretches nor shrinks the segment.

To translate this condition on a rigid motion of the framework into a condition on its velocity vectors, we must take a closer look at the structure of these velocity vectors. The velocity vector at an endpoint can be resolved into a vector perpendicular to the segment and a vector parallel to the segment.

In Figure 3.5, we illustrate this resolution with the first and third motions from the previous figure. The vectors of the second motion resolve into themselves (perpendicular to the segment) and the zero vector (parallel to the segment).

Recall that the motion under consideration was given by $\mathbf{P}_1, \mathbf{P}_2$, where $\mathbf{P}_i(t) = \big(x_i(t), y_i(t)\big)$. So we have $\big(x_2(t) - x_1(t)\big)^2 + \big(y_2(t) - y_1(t)\big)^2 = h^2$, where $h > 0$ is the length of the segment. Differentiating each side with respect to t gives the condition

$$2\big(x_2(t) - x_1(t)\big)\big(x_2'(t) - x_1'(t)\big) + 2\big(y_2(t) - y_1(t)\big)\big(y_2'(t) - y_1'(t)\big) = 0;$$

which can be rewritten as the inner product

$$\big(x_2(t) - x_1(t), y_2(t) - y_1(t)\big) \cdot \big(x_2'(t) - x_1'(t), y_2'(t) - y_1'(t)\big) = 0.$$

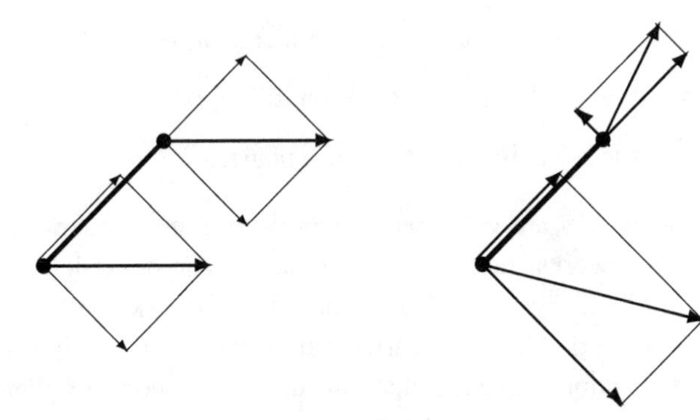

FIGURE 3.5

Evaluating the terms in this equation at time $\overline{t}$ gives

$$(\overline{x}_2 - \overline{x}_1, \overline{y}_2 - \overline{y}_1) \cdot (u_2 - u_1, v_2 - v_1) = 0.$$

To interpret this equation in terms of the velocity vectors, we now rewrite it as

$$\frac{(\overline{x}_2 - \overline{x}_1, \overline{y}_2 - \overline{y}_1)}{h} \cdot (u_2, v_2) = \frac{(\overline{x}_2 - \overline{x}_1, \overline{y}_2 - \overline{y}_1)}{h} \cdot (u_1, v_1).$$

Note that the absolute value of the inner product

$$\frac{(\overline{x}_2 - \overline{x}_1, \overline{y}_2 - \overline{y}_1)}{h} \cdot (u_i, v_i)$$

is the length of the projection of the velocity vector (u_i, v_i) onto the vector $(\overline{x}_2 - \overline{x}_1, \overline{y}_2 - \overline{y}_1)$, and its sign gives the direction of this projection. Hence:

The inner product condition requires that the parallel projections of the two velocity vectors have the same length and direction.

We are now almost ready to define *infinitesimal rigidity,* which, like rigidity, has a common definition in all dimensions. However, for technical reasons (explained later), we must differentiate between two classes of frameworks, *constricted* and *normal.* The problem frameworks are those that are "constricted" into some linear space (point, line or plane) of dimension lower than the dimension one would expect the framework to occupy.

Let $\mathcal{F} = (V, E, \mathbf{p})$ be an m-dimensional framework, let $|V| = n$ and let $P = \{\mathbf{p}_1, \ldots, \mathbf{p}_n\}$. We say that $\mathcal{F}$ is *constricted* if

- $m = 1, 2, 3$, $n > 1$ but all points in P are identical; or
- $m = 2, 3$, $n > 2$ and all points in P lie on a line; or
- $m = 3$, $n > 3$ and all points in P lie on a plane.

Constricted frameworks are very special, and we will use the term *normal* to denote those frameworks that are not constricted. The definition we will give for the infinitesimal rigid motions of normal frameworks will not work for constricted frameworks. Since constricted frameworks are really not very interesting, we will simply declare that, by definition, none of them are infinitesimally rigid. We leave development of constricted frameworks and the proof of this fact to the interested reader through a series of exercises.

Note that certain small frameworks are not constricted even though they lie on a lower dimensional space. For instance, a framework consisting of a single point is still normal in every dimension; so is a single segment, as long as it has positive length. Also a triangle whose vertices do not all lie on a line is normal in 2- and 3-space.

For normal frameworks the definition of infinitesimal rigidity is parallel to the definition of rigidity and, like rigidity, has a common definition for all dimensions. (The definition of an infinitesimal motion of a framework works for all frameworks and is stated in that generality.) We use these terms:

- An *infinitesimal motion* of the m-dimensional framework $(V, E, \mathbf{p})$ is a function $\mathbf{q} : V \to R^m$ so that $(\mathbf{p}_i - \mathbf{p}_j) \cdot (\mathbf{q}_i - \mathbf{q}_j) = 0$, for all $\{a_i, a_j\} \in E$.
- An infinitesimal motion $\mathbf{q}$ of the normal framework $(V, E, \mathbf{p})$ is called an *infinitesimal rigid motion* if $(\mathbf{p}_i - \mathbf{p}_j) \cdot (\mathbf{q}_i - \mathbf{q}_j) = 0$, for all $a_i, a_j \in V$.
- An infinitesimal motion $\mathbf{q}$ of the normal framework $(V, E, \mathbf{p})$ is an *infinitesimal deformation* if $(\mathbf{p}_i - \mathbf{p}_j) \cdot (\mathbf{q}_i - \mathbf{q}_j) \neq 0$, for some $a_i, a_j \in V$.
- A normal framework $(V, E, \mathbf{p})$ is said to be *infinitesimally rigid* if all of its infinitesimal motions are infinitesimal rigid motions.

To illustrate this definition, let's revisit the concrete examples introduced in Section 2.2 concerning the (normal) framework $\mathcal{F} = (V, E, \mathbf{p})$, where $V = \{a_1, a_2, a_3\}$, $E = \{\{a_1, a_2\}, \{a_2, a_3\}\}$ $\mathbf{p} : V \to R^2$ is given by

$$\mathbf{p}(a_1) = \mathbf{p}_1 = (-6, 0),$$

$$\mathbf{p}(a_2) = \mathbf{p}_2 = (0, 0),$$

$$\mathbf{p}(a_3) = \mathbf{p}_3 = (-4, 2).$$

In that section, we introduced the rigid motion

$$\mathbf{P}_1(t) = \left(t - 3\sqrt{4 - t^2}, 5t - \frac{t^2}{2}\right);$$

$$\mathbf{P}_2(t) = \left(t, 2t - \frac{t^2}{2}\right);$$

$$\mathbf{P}_3(t) = \left(2t - 2\sqrt{4 - t^2}, 4t - \frac{t^2}{2} + \sqrt{4 - t^2}\right).$$

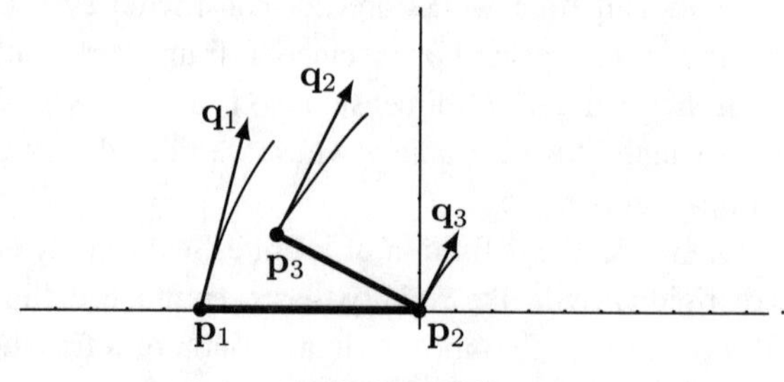

FIGURE 3.6

By differentiating these functions with respect to t and evaluating them at $t = 0$, we obtain the initial velocity vectors for this motion, which are pictured in Figure 3.6:

$$\mathbf{P}_1'(t) = \left(1 + \frac{3t}{\sqrt{4-t^2}},\ 5 - t\right),$$

$$\mathbf{P}_2'(t) = (1, 2 - t),$$

$$\mathbf{P}_3'(t) = \left(2 + \frac{2t}{\sqrt{4-t^2}}, 4 - t - \frac{t}{\sqrt{4-t^2}}\right)$$

and

$$\mathbf{q}_1 = \mathbf{P}_1'(0) = (1, 5),$$

$$\mathbf{q}_2 = \mathbf{P}_2'(0) = (1, 2),$$

$$\mathbf{q}_3 = \mathbf{P}_3'(0) = (2, 4).$$

We verify that these vectors describe an infinitesimal motion of $\mathcal{F}$ by computing

$$(\mathbf{p}_1 - \mathbf{p}_2) \cdot (\mathbf{q}_1 - \mathbf{q}_2) = \big((-6, 0) - (0, 0)\big) \cdot \big((1, 5) - (1, 2)\big) = 0,$$

$$(\mathbf{p}_3 - \mathbf{p}_2) \cdot (\mathbf{q}_3 - \mathbf{q}_2) = \big((-4, 2) - (0, 0)\big) \cdot \big((2, 4) - (1, 2)\big) = 0.$$

We then show that this infinitesimal motion is an infinitesimal rigid motion by computing

$$(\mathbf{p}_3 - \mathbf{p}_1) \cdot (\mathbf{q}_3 - \mathbf{q}_1) = \big((-4, 2) - (-6, 0)\big) \cdot \big((2, 4) - (1, 5)\big) = 0.$$

Recall that we considered a second example by changing the third function $\mathbf{P}_3(t)$ to

$$\mathbf{P}_3(t) = \left(t\sqrt{4-t^2} + 2t^2 + t - 4, 2t\sqrt{4-t^2} - \frac{3t^2}{2} + 2t + 2\right)$$

while retaining $\mathbf{P}_1(t)$ and $\mathbf{P}_2(t)$ unchanged. This results in another infinitesimal motion of $\mathcal{F}$, in this case an infinitesimal deformation.

Exercise 3.1. *Consider $\mathbf{P}_1(t)$, $\mathbf{P}_2(t)$ and the new $\mathbf{P}_3(t)$.*

1. *Compute the new $\mathbf{q}_3$.*
2. *Show that $\mathbf{q}_1$, $\mathbf{q}_2$ and $\mathbf{q}_3$ form an infinitesimal motion of $\mathcal{F}$.*
3. *Show that $\mathbf{q}_1$, $\mathbf{q}_2$ and $\mathbf{q}_3$ form an infinitesimal deformation of $\mathcal{F}$.*

As an example of the problem posed by constricted frameworks, re-embed this framework as a constricted framework: Let $\mathbf{p}_1 = (-6, 0)$ and $\mathbf{p}_2 = (0, 0)$, as before, but let $\mathbf{p}_3 = (\sqrt{20}, 0)$. See Figure 3.7.

Of course this framework is not rigid: Keep the left-hand segment fixed and rotate the right-hand framework counterclockwise around $\mathbf{p}_2$. Differentiating this motion yields the infinitesimal motion $\mathbf{q}_1 = \mathbf{q}_2 = (0, 0)$ and $\mathbf{q}_3 = (0, a)$, where a is a positive number depending on the rate of rotation. Since

$$(\mathbf{p}_1 - \mathbf{p}_2) \cdot (\mathbf{q}_1 - \mathbf{q}_2) = (\mathbf{p}_3 - \mathbf{p}_2) \cdot (\mathbf{q}_3 - \mathbf{q}_2) = 0,$$

this is an infinitesimal motion. But we also have

$$(\mathbf{p}_3 - \mathbf{p}_1) \cdot (\mathbf{q}_3 - \mathbf{q}_1) = 0!$$

So if we were to apply the definition we have for normal frameworks, we would conclude that this infinitesimal motion is an infinitesimal rigid motion for this framework. In fact, using that definition, we could prove that this framework is infinitesimally rigid. We have a similar problem with most constricted frameworks.

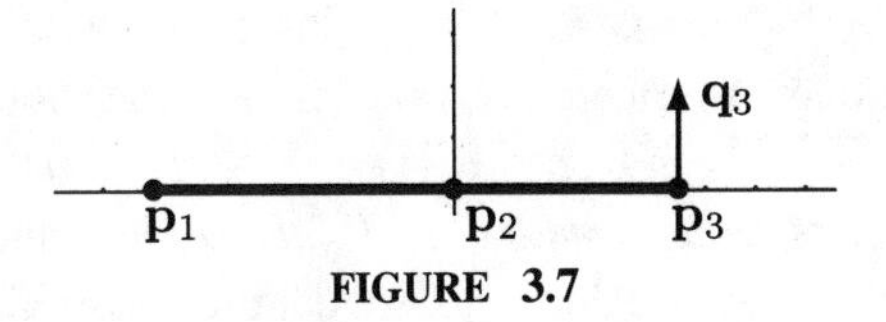

FIGURE 3.7

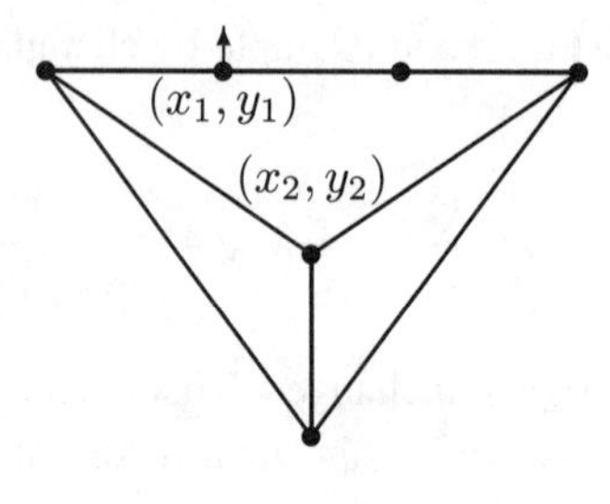

FIGURE 3.8

Now that we have introduced two notions of rigidity, it is natural to ask:

Are rigidity and infinitesimal rigidity really different?

The answer is "Yes, they are different," as a 2-dimensional framework we discussed earlier shows. This framework consists of a quadrilateral with diagonal that is rigid and a "chain" of three segments stretched taut; see Figure 3.8.

As we observed earlier, this framework is rigid. But it is not infinitesimally rigid: Let $\mathbf{q}$ assign the zero vector to every point except the point labeled (x_1, y_1), and to this point let $\mathbf{q}$ assign a nonzero vector perpendicular to the "chain." Since the projections of the vector assigned to the endpoints of each segment are zero, $\mathbf{q}$ does not distort any of the segments in the framework, but it does distort the distance from (x_1, y_1) to (x_2, y_2). Therefore, $\mathbf{q}$ is an infinitesimal deformation. One might think of this infinitesimal motion as a vibration of the chain.

In view of this example, we now ask:

Can a framework be infinitesimally rigid but not rigid?

Here, the answer is "No."

Theorem 3.1. *If the m-dimensional framework $(V, E, \mathbf{p})$ is infinitesimally rigid, then it is rigid.*

A complete proof of this result involves some advanced arguments to handle special cases and is beyond the scope of this book. Nevertheless, we sketch the main steps of the proof, which are accessible. Actually we outline a proof of the contrapositive: If $(V, E, \mathbf{p})$ is not rigid, then it is not infinitesimally rigid. Also, to simplify the notation, we restrict this sketch to the 2-dimensional case; but the techniques used here are valid in all dimensions.

Let $V = \{a_1, \ldots, a_n\}$ and let $\{\mathbf{P}_i(t) : 1 \leq i \leq n\}$ be a motion of this framework that is a deformation. If $(V, E, \mathbf{p})$ is constricted, it is not infinitesimally rigid and we are done. So, we continue assuming that $(V, E, \mathbf{p})$

is normal. For the vertex $a_i \in V$, let $\mathbf{P}_i(t) = \big(x_i(t), y_i(t)\big)$ denote the trajectory of the point $\mathbf{p}_i$ under the motion and let $h_{ij}(t)$ denote the distance between $\mathbf{P}_i(t)$ and $\mathbf{P}_j(t)$. Differentiating both sides of

$$\big(x_i(t) - x_j(t)\big)^2 + \big(y_i(t) - y_j(t)\big)^2 = h_{ij}^2(t)$$

gives

$$2\big(x_i(t) - x_j(t)\big)\big(x_i'(t) - x_j'(t)\big) + 2\big(y_i(t) - y_j(t)\big)\big(y_i'(t) - y_j'(t)\big) = 2h_{ij}(t)h_{ij}'(t).$$

If $(a_i, a_j) \in E$, $h_{ij}(t)$ is a constant and $h_{ij}'(t) = 0$, for all t. Thus, in this case,

$$\big(x_i(0) - x_j(0), y_i(0) - y_j(0)\big) \cdot \big(x_i'(0) - x_j'(0), y_i'(0) - y_j'(0)\big) = h_{ij}(0)h_{ij}'(0) = 0;$$

and we conclude that $\mathbf{q} : V \to R^2$ defined by $\mathbf{q}_i = \big(x_i'(0), y_i'(0)\big)$ is an infinitesimal motion of $(V, E, \mathbf{p})$. We wish to show that $\mathbf{q}$ is an infinitesimal deformation of $(V, E, \mathbf{p})$.

Suppose then that $\mathbf{p}_i$ and $\mathbf{p}_j$ are distinct, that $(a_i, a_j) \notin E$ and that $h_{ij}(t)$ is not a constant. If $h_{ij}'(0) \neq 0$, then

$$\big(x_i(0) - x_j(0), y_i(0) - y_j(0)\big) \cdot \big(x_i'(0) - x_j'(0), y_i'(0) - y_j'(0)\big) = h_{ij}(0)h_{ij}'(0) \neq 0$$

and $\mathbf{q}$ is an infinitesimal deformation.

Of course, $h_{ij}(t)$ could be a constant for all i and j and all $t \in [0, a]$. This problem can be avoided by constructing a new motion that starts the deformation at once. However, even if $h_{ij}(t)$ is not a constant in any initial interval, $h_{ij}'(0)$ could still be 0. Dealing with this case is more complicated.

Exercise 3.2. *Consider* 1*-dimensional rigidity.*

1. *Prove the following restricted converse to Theorem 3.1: Let $\mathcal{F} = (V, E, \mathbf{p})$ be a* 1*-dimensional framework, where $\mathbf{p}$ is one-to-one. If $\mathcal{F}$ is rigid, then it is infinitesimally rigid.*
2. *Explain why the condition "where $\mathbf{p}$ is one-to-one" is necessary.*

3.2 Infinitesimal Rigidity

We started the last section with an example that illustrated just how difficult it is to work with the quadratic equations of rigidity. We revisit that example to illustrate the relative ease in working with the linear equations of infinitesimal rigidity.

Recall the framework $(V, E, \mathbf{p})$, in which V is the set of four vertices $\{a_1, a_2, a_3, a_4\}$; E the five edges $\{a_1, a_2\}$, $\{a_2, a_3\}$, $\{a_3, a_4\}$, $\{a_1, a_4\}$ and $\{a_1, a_3\}$; and $\mathbf{p}_1 = (0, 0)$, $\mathbf{p}_2 = (3, 0)$, $\mathbf{p}_3 = (3, 2)$ and $\mathbf{p}_4 = (0, 2)$. We have redrawn this framework in Figure 3.9. Notice that it now includes an infinitesimal motion $\mathbf{q} : V \to R^2$, where $\mathbf{q}_i = (u_i, v_i)$, for $i = 1, \ldots, 4$.

Each segment of the framework gives a linear equation that must be satisfied by the coordinates of the vectors of the infinitesimal motion. For example, the linear equation for the bottom segment is

$$(3 - 0, 0 - 0) \cdot (u_2 - u_1, v_2 - v_1) = 0$$

or simply

$$3(u_2 - u_1) = 0.$$

The five segments together give the system

$$\begin{aligned} 3(u_2 - u_1) & & &= 0; \\ & & 2(v_4 - v_1) &= 0; \\ 3(u_4 - u_3) & & &= 0; \\ & & 2(v_3 - v_2) &= 0; \\ 3(u_3 - u_1) &+ & 2(v_3 - v_1) &= 0. \end{aligned}$$

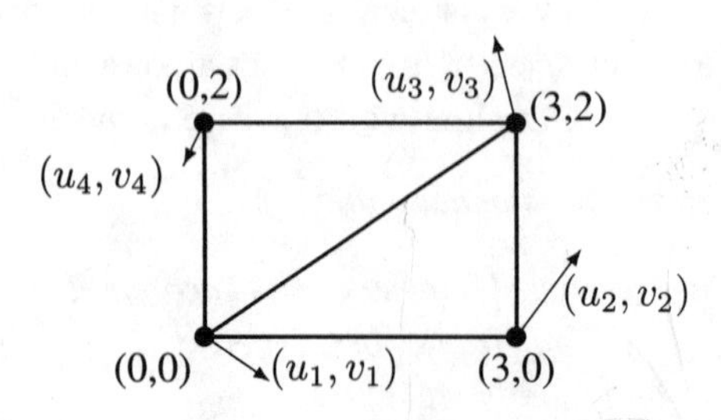

FIGURE 3.9

We wish to prove that this normal framework is infinitesimally rigid. To accomplish this, we must show that every infinitesimal motion of this framework is an infinitesimal rigid motion. To do that, we must show that the condition that the segment between $(0, 2)$ and $(3, 0)$ is not being stretched or compressed by this infinitesimal motion; that is, we must show that the linear equation

$$3(u_4 - u_2) - 2(v_4 - v_2) = 0,$$

follows from the above system of linear equations.

From the first four equations, we see that $u_1 = u_2$, $u_3 = u_4$, $v_1 = v_4$ and $v_3 = v_2$. Substituting these values for u_1, u_3, v_1 and v_3 in the fifth equation, gives the equation we seek. Done!

Comparing this infinitesimal rigidity derivation with the corresponding derivation using the quadratic equations of rigidity, it is clear that it is much, much easier to work with infinitesimal rigidity. Not only is infinitesimal rigidity much easier to work with but, in view of Theorem 3.1, requiring that our structures be infinitesimally rigid is not a safety risk. In fact, as illustrated by the example in Figure 3.8, infinitesimal rigidity is a more robust form of rigidity. You really would want your structures to be infinitesimally rigid instead of being simply rigid.

We must now build a "mathematical framework" within which to study infinitesimal motions. Let $(V, E, \mathbf{p})$ denote an arbitrary but fixed framework in m-dimensional space, and let $V = \{a_1, \ldots, a_n\}$. Depending on the value of m, let x_i, (x_i, y_i) or (x_i, y_i, z_i) denote the coordinates of $\mathbf{p}_i$, for $i = 1, \ldots, n$. Next let $\mathbf{q} : V \to R^m$ be an arbitrary function from the vertex set into the vector space R^m and, again depending on m, let u_i, (u_i, v_i) or (u_i, v_i, w_i) denote the coordinates of $\mathbf{q}_i$, for $i = 1, \ldots, n$. We think of the vector $\mathbf{q}_i$ as being assigned to the point $\mathbf{p}_i$.

There is a natural way of identifying this function $\mathbf{q}$ with a single vector in R^{mn}, namely, the vector obtained by stringing together, in order, the vectors assigned by $\mathbf{q}$:

- Identify $\mathbf{q}$ with $(u_1, \ldots, u_n) \in R^n$, if $m = 1$.
- Identify $\mathbf{q}$ with $(u_1, v_1, \ldots, u_n, v_n) \in R^{2n}$, if $m = 2$.
- identify $\mathbf{q}$ with $(u_1, v_1, w_1 \ldots, u_n, v_n, w_n) \in R^{3n}$, if $m = 3$.

Thus the set of all functions from V into R^m may be identified with the vector space R^{mn}. Furthermore, the interested reader may easily check that scalar multiplication and addition of these functions correspond to scalar multiplication and addition in the vector space.

Suppose that the framework $(V, E, \mathbf{p})$ included the segment joining $\mathbf{p}_i$ and $\mathbf{p}_j$ ($\{a_i, a_j\} \in E$). Then, if $\mathbf{q}$ were an infinitesimal motion of this framework, $\mathbf{q}$ must satisfy

$$(x_i - x_j) \cdot (u_i - u_j) = 0, \quad \text{if } m = 1;$$

$$(x_i - x_j, y_i - y_j) \cdot (u_i - u_j, v_i - v_j) = 0, \quad \text{if } m = 2;$$

$$(x_i - x_j, y_i - y_j, z_i - z_j) \cdot (u_i - u_j, v_i - v_j, w_i - w_j) = 0, \quad \text{if } m = 3;$$

Equivalently, the vector in R^{mn} corresponding to $\mathbf{q}$ must satisfy the linear equation (remember the x_i, y_i and z_i are all fixed constants)

$$(x_i - x_j)u_i - (x_i - x_j)u_j = 0, \quad \text{if } m = 1;$$

$$(x_i - x_j)u_i + (y_i - y_j)v_i - (x_i - x_j)u_j - (y_i - y_j)v_j = 0, \quad \text{if } m = 2;$$

$$\begin{aligned} &(x_i - x_j)u_i + (y_i - y_j)v_i + (z_i - z_j)w_i \\ &\quad -(x_i - x_j)u_j - (y_i - y_j)v_j - (z_i - z_j)w_j = 0, \quad \text{if } m = 3. \end{aligned}$$

Thus, corresponding to each edge in E, there is a linear equation that must be satisfied by the coordinates of the vector associated with any infinitesimal motion $\mathbf{q}$. It follows that the set of vectors in R^{mn} that corresponds to the infinitesimal motions of this framework is the solution set to the collection of linear equations corresponding to the edges in E. Also, the set of infinitesimal rigid motions of this framework is the solution set to the collection of linear equations corresponding to all pairs of distinct vertices in V. Since each of these equations is homogeneous, the above solution sets are both subspaces.

Denote by $\mathcal{M}(\mathcal{F})$ *the subspace of infinitesimal motions of the framework* $\mathcal{F} = (V, E, \mathbf{p})$. If $\mathcal{F}$ is normal, denote by $\mathcal{R}(\mathcal{F})$ *the subspace of infinitesimal rigid motions.* Since every infinitesimal rigid motion of $\mathcal{F}$ is an infinitesimal motion of $\mathcal{F}$,

$$\mathcal{R}(\mathcal{F}) \subseteq \mathcal{M}(\mathcal{F}).$$

In terms of these subspaces, $\mathcal{F}$ is infinitesimally rigid if and only if $\mathcal{R}(\mathcal{F}) = \mathcal{M}(\mathcal{F})$. And since $\mathcal{R}(\mathcal{F}) \subseteq \mathcal{M}(\mathcal{F})$, $\mathcal{F}$ is infinitesimally rigid if and only if

$$\dim[\mathcal{R}(\mathcal{F})] = \dim[\mathcal{M}(\mathcal{F})].$$

Since we will frequently refer to this characterization of infinitesimal rigidity, we formalize it:

Lemma 3.2. *Let $\mathcal{F} = (V, E, \mathbf{p})$ be a normal m-dimensional framework. Then*

$$\dim[\mathcal{R}(\mathcal{F})] \leq \dim[\mathcal{M}(\mathcal{F})]$$

and $\mathcal{F}$ is infinitesimally rigid if and only if

$$\dim[\mathcal{R}(\mathcal{F})] = \dim[\mathcal{M}(\mathcal{F})].$$

We illustrate these definitions and Lemma 3.2 with a simple example of a rectangle in the plane. Take as $\mathcal{F} = (V, E, \mathbf{p})$ the rectangle in Figure 3.9 with the diagonal removed. The subspace of infinitesimal motions, $\mathcal{M}(\mathcal{F})$, is the solution set of the system

$$\begin{array}{cccccccccc}
-3u_1 & & +3u_2 & & & & & & = & 0 \\
& -2v_1 & & & & & & +2v_4 & = & 0 \\
& & & & -3u_3 & & +3u_4 & & = & 0 \\
& & & -2v_2 & & +2v_3 & & & = & 0
\end{array}$$

and the space of infinitesimal rigid motions, $\mathcal{R}(\mathcal{F})$, is the solution set of the system:

$$\begin{array}{cccccccccc}
-3u_1 & & +3u_2 & & & & & & = & 0 \\
& -2v_1 & & & & & & +2v_4 & = & 0 \\
& & & & -3u_3 & & +3u_4 & & = & 0 \\
& & & -2v_2 & & +2v_3 & & & = & 0 \\
-3u_1 & -2v_1 & & & +3u_3 & +2v_3 & & & = & 0 \\
& & 3u_2 & -2v_2 & & & -3u_4 & +2v_4 & = & 0
\end{array}$$

Applying Gauss–Jordan row reduction to each system gives

$$\begin{array}{cccccccccc}
u_1 & & -u_2 & & & & & & = & 0 \\
& v_1 & & & & & & -v_4 & = & 0 \\
& & & v_2 & & -v_3 & & & = & 0 \\
& & & & u_3 & & -u_4 & & = & 0
\end{array}$$

and

$$\begin{array}{cccccccccccc} u_1 & & & & & -\frac{2}{3}v_3 & -u_4 & +\frac{2}{3}v_4 & = & 0 \\ & v_1 & & & & & & -v_4 & = & 0 \\ & & u_2 & & & -\frac{2}{3}v_3 & -u_4 & +\frac{2}{3}v_4 & = & 0 \\ & & & v_2 & & -v_3 & & & = & 0 \\ & & & & u_3 & & -u_4 & & = & 0 \end{array}$$

Note that the second system was redundant and one equation has dropped out.

We may now see that $\mathcal{M}(\mathcal{F})$ is 4-dimensional and consists of vectors of the form (a, d, a, b, c, b, c, d), for arbitrary $a, b, c, d \in R$, while $\mathcal{R}(\mathcal{F})$ is 3-dimensional and consists of vectors of the form

$$\left(\tfrac{2}{3}b + c - \tfrac{2}{3}d, d, \tfrac{2}{3}b + c - \tfrac{2}{3}d, b, c, b, c, d\right),$$

for arbitrary $b, c, d \in R$. Clearly, $\mathcal{R}(\mathcal{F})$ is a subspace of $\mathcal{M}(\mathcal{F})$. Specifically, it is the subspace of those vectors in $\mathcal{M}(\mathcal{F})$, as just described, that satisfy the additional condition that

$$a = \tfrac{2}{3}b + c - \tfrac{2}{3}d.$$

It is also clear that $\mathcal{R}(\mathcal{F})$ is a proper subspace of $\mathcal{M}(\mathcal{F})$. These observations correspond to what we already know: the rectangle without diagonals is not rigid and hence, by Theorem 3.1, is not infinitesimally rigid.

To tie this all together, let's give a direct proof that this framework is not infinitesimally rigid by selecting a vector assignment in $\mathcal{M}(\mathcal{F})$ but not in $\mathcal{R}(\mathcal{F})$. To do this, we select a, b, c and d that do not satisfy $a = \frac{2}{3}b + c - \frac{2}{3}d$;

$$a = b = c = \tfrac{1}{2}, \quad d = -\tfrac{1}{2}$$

will do. This gives the vector $\frac{1}{2}(1, -1, 1, 1, 1, 1, 1, -1)$, which corresponds to the vector assignment:

$$\begin{array}{ll} (u_1, v_1) = \left(\frac{1}{2}, -\frac{1}{2}\right), & (u_2, v_2) = \left(\frac{1}{2}, \frac{1}{2}\right), \\ (u_3, v_3) = \left(\frac{1}{2}, \frac{1}{2}\right), & (u_4, v_4) = \left(\frac{1}{2}, -\frac{1}{2}\right). \end{array}$$

This infinitesimal motion is pictured in Figure 3.10. One easily verifies directly that it is an infinitesimal motion of the framework but not an infinitesimal rigid motion.

Exercise 3.1. *Consider each of the following vector assignments for the framework pictured in Figure 3.10.*

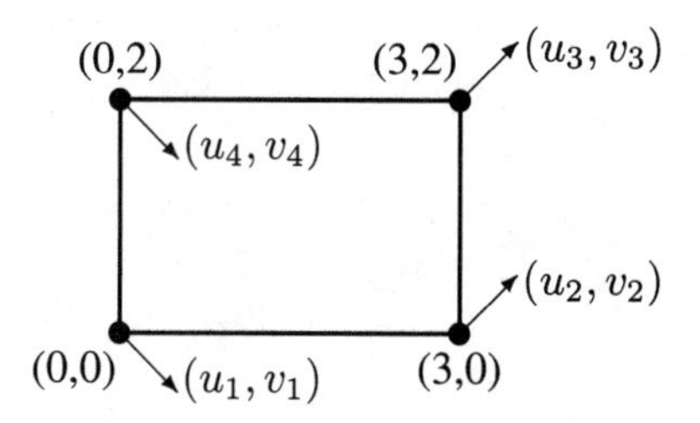

FIGURE 3.10

1. *In each case, sketch the vector assignment.*
2. *For each assignment, speculate whether or not it is: a motion of the framework, a rigid motion of the framework, a deformation of the framework.*
3. *Use the equations to check your speculations.*

   (a) $(u_1, v_1) = (1,4)$, $(u_2, v_2) = (-3,-2)$, $(u_3, v_3) = (2,-3)$, and $(u_4, v_4) = (2,3)$.

   (b) $(u_1, v_1) = (0,1)$, $(u_2, v_2) = (0,-2)$, $(u_3, v_3) = (2,-2)$, and $(u_4, v_4) = (2,1)$.

   (c) $(u_1, v_1) = (2,-1)$, $(u_2, v_2) = (2,-1)$, $(u_3, v_3) = (3,-1)$, and $(u_4, v_4) = (3,-1)$.

   (d) $(u_1, v_1) = (-2,3)$, $(u_2, v_2) = (-2,-3)$, $(u_3, v_3) = (3,-3)$, and $(u_4, v_4) = (3,3)$.

   (e) $(u_1, v_1) = (-1,1)$, $(u_2, v_2) = (-1,-2)$, $(u_3, v_3) = (1,-2)$, and $(u_4, v_4) = (1,1)$.

Exercise 3.2. *Consider the* 1*-dimensional framework* $\mathcal{F} = (V, E, \mathbf{p})$ *where* $V = \{a_1, a_2, a_3\}$, E *consist of the single edge* $\{a_1, a_2\}$ *and* $\mathbf{p} : V \to R^1$ *is given by* $\mathbf{p}_1 = -3$, $\mathbf{p}_2 = 7$ *and* $\mathbf{p}_3 = 2$.

1. *Give the systems of equations that determine* $\mathcal{M}(\mathcal{F})$ *and* $\mathcal{R}(\mathcal{F})$.
2. *Find an expression for the typical vector in each of these spaces.*

Let's consider 1-dimensional infinitesimal rigidity in more detail. Let $\mathcal{F} = (V, E, \mathbf{p})$ be a 1-dimensional framework, where $V = \{a_1, \ldots, a_n\}$ and where the points $\mathbf{p}_i$ are distinct, for all $i = 1, \ldots, n$. Note that this condition implies that the framework is normal. Let the coordinate of $\mathbf{p}_i$ be x_i, and let $(u_1, \ldots, u_n)$ be the n-dimensional vector associated with the infinitesimal motion $\mathbf{q}$. The equation associated with the edge $\{a_i, a_j\} \in E$ is simply

$$(x_i - x_j)u_i - (x_i - x_j)u_j = 0.$$

Since $x_i \neq x_j$, the only solution is $u_i = u_j$. It follows that, if a_i and a_j are joined by a path $a_i = a_{i_1}, \ldots, a_{i_k} = a_j$ in (V, E), then $u_i = u_{i_1} = \cdots = u_{i_k} = u_j$. We conclude that $\mathbf{q}$ is constant on the sets of vertices of the components of (V, E). It is also easy to see that any $\mathbf{q} : V \to R^1$ that is constant on the sets of vertices of the components of (V, E) is an infinitesimal motion of $\mathcal{F}$. Thus $\mathcal{M}(\mathcal{F})$ is the space of functions that are constant on the vertices of the components of (V, E), and $\dim[\mathcal{M}(\mathcal{F})]$ is simply the number of components of (V, E). On the other hand, one easily sees that $\mathcal{R}(\mathcal{F})$ is the space of constant functions on V and $\dim[\mathcal{R}(\mathcal{F})] = 1$.

We have proved the infinitesimal analogue to Theorem 2.16:

Theorem 3.3. *A 1-dimensional framework $(V, E, \mathbf{p})$, where $\mathbf{p}$ is one-to-one, is infinitesimally rigid if and only if the graph (V, E) is connected.*

Corollary 3.4. *The concepts of rigidity and infinitesimal rigidity coincide for almost all 1-dimensional frameworks, that is, for all frameworks $(V, E, \mathbf{p})$, where $\mathbf{p}$ is one-to-one.*

Exercise 3.3. *Consider a slight generalization of Theorem 3.3:*

A 1-dimensional framework $(V, E, \mathbf{p})$, where $\mathbf{p}_i \neq \mathbf{p}_j$ whenever $\{a_i, a_j\} \in E$, is infinitesimally rigid if and only if the graph (V, E) is connected.

1. *Prove this generalization.*
2. *Explain what would "go wrong" if $\mathbf{p}_i = \mathbf{p}_j$ for some $\{a_i, a_j\} \in E$.*

The translations of the space R^1 are the functions $T_u : R \to R$ by $T_u(x) = x + u$. If we think of this translation as developing over time, we write it as a function of two variables,

$$T_u(x, t) = x + ut,$$

where $x \in R$ and $t \in [0, 1]$. For the point with coordinate x, $T(x, 0) = x$ is its initial position, and $T(x, 1) = x + u$ is its final position under the translation. Differentiating with respect to time and evaluating at $t = 0$, we have

$$T_u'(x) = u.$$

Thus the infinitesimal translation $\tau_u(x)$ associated with the translation $T_u(x)$ is the constant function that assigns u to each point on the line. We conclude that the space of infinitesimal rigid motions, $\mathcal{R}(\mathcal{F})$, of any linear framework $\mathcal{F} = (V, E, \mathbf{p})$ is simply the 1-dimensional space of constant functions, or infinitesimal translations, restricted to the vertices of that framework.

The situation is similar in all higher dimensions. Let $\mathcal{F} = (V, E, \mathbf{p})$ be a framework in R^m; then each vector in $\mathcal{R}(\mathcal{F})$ corresponds to the restriction of an infinitesimal rigid motion of R^m to the points of the framework. We will prove this in dimensions two and three as we go along. But having this in mind now helps motivate our development.

The trajectories of the points of the plane under a translation of the plane are parallel segments, and all points move along their segments at the same rate. Hence an *infinitesimal translation* of the plane assigns the same vector to each point in the plane. Let $\tau_{(u,v)} : R^2 \to R^2$ be the infinitesimal translation that assigns the vector (u, v) to each point in the plane, that is, the constant function $\tau_{(u,v)}(x, y) = (u, v)$. If (x, y) and $(\hat{x}, \hat{y})$ are any two points in the plane,

$$\begin{aligned}(\hat{x} - x, \hat{y} - y) \cdot \big(\tau_{(u,v)}(\hat{x}, \hat{y}) - \tau_{(u,v)}(x, y)\big) &= (\hat{x} - x, \hat{y} - y) \cdot (0, 0) \\ &= 0,\end{aligned}$$

confirming that the infinitesimal translations neither shrink nor stretch the distance between any pair of points in the plane. In other words, it is an infinitesimal rigid motion of the plane. The collection of all infinitesimal translations forms a 2-dimensional vector space spanned by $\tau_{(1,0)}$ and $\tau_{(0,1)}$: For any infinitesimal translation $\tau_{(u,v)}$, we have

$$\tau_{(u,v)} = u\tau_{(1,0)} + v\tau_{(0,1)}.$$

Next we consider infinitesimal rotations. To motivate the definition, consider an actual rotation of the plane about the origin. Let (x, y) be any point in the plane, and rewrite it in terms of its polar coordinates

$$(r, \theta) : (x, y) = \big(r \sin(\theta), r \cos(\theta)\big).$$

The trajectory of (x, y) under a rotation about the origin is given by

$$\big(x(t), y(t)\big) = \big(r \sin(\theta + at), r \cos(\theta + at)\big),$$

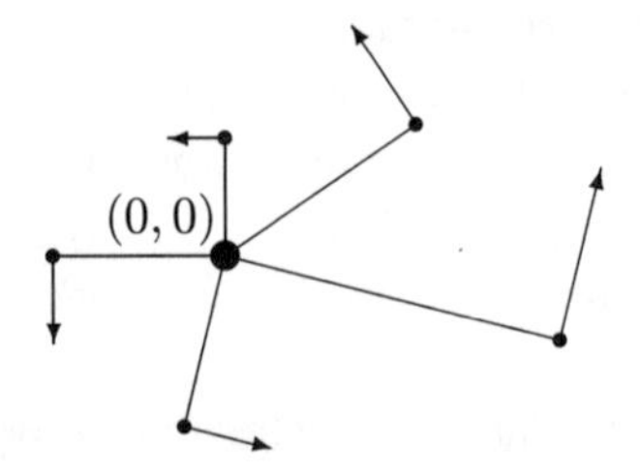

FIGURE 3.11

for time $t \in [0,1]$. The rotation is counterclockwise when a is positive, and the rotation is clockwise when a is negative. The larger the absolute value of a, the faster the rotation. Differentiating with respect to time and evaluating the derivative at $t = 0$, we have

$$a\big(r\cos(\theta), -r\sin(\theta)\big) = a(y, -x).$$

Thus the initial velocity vector of a rotation at a point (x, y) is perpendicular to the segment from that point to the center of rotation, and its length is proportional to the distance from that point to that center. See Figure 3.11.

Taking these observations into account, we define the *unit infinitesimal rotation* with center (x_0, y_0), $\rho_{(x_0,y_0)} : R^2 \to R^2$ by

$$\rho_{(x_0,y_0)}(x, y) = (y - y_0, x_0 - x).$$

The remaining infinitesimal rotations with center (x_0, y_0) are simply the scalar multiples of $\rho_{(x_0,y_0)}$. To verify that $\rho_{(x_0,y_0)}$ is indeed an infinitesimal rigid motion of the entire plane, let (x, y) and $(\hat{x}, \hat{y})$ be any two points in the plane and compute

$$\begin{aligned}
(\hat{x} - x, \hat{y} - y) \cdot \big(\rho_{(x_0,y_0)}(\hat{x}, \hat{y}) - \rho_{(x_0,y_0)}(x, y)\big) \\
&= (\hat{x} - x, \hat{y} - y) \cdot \big((\hat{y} - y_0, x_0 - \hat{x}) - (y - y_0, x_0 - x)\big) \\
&= (\hat{x} - x, \hat{y} - y) \cdot (\hat{y} - y, x - \hat{x})) \\
&= 0.
\end{aligned}$$

The next lemma summarizes our investigation of infinitesimal rigid motions of the plane.

Lemma 3.5. *The set of all infinitesimal translations and rotations of the plane forms a 3-dimensional vector space.*

Proof. Clearly, $\tau_{(1,0)}$, $\tau_{(0,1)}$ and $\rho_{(0,0)}$ are independent infinitesimal motions of the plane. We have already noted that any infinitesimal translation of the plane is a linear combination of $\tau_{(1,0)}$ and $\tau_{(0,1)}$. The next equation, which is easily checked by evaluating each side at an arbitrary point (x, y), shows that any infinitesimal rotation of the plane is a linear combination of these three vectors:

$$a\rho_{(x_0,y_0)} = a\rho_{(0,0)} + (-ay_0)\tau_{(1,0)} + (ax_0)\tau_{(0,1)}.$$

Finally, this same equation shows that any linear combination of these three vectors that includes $\rho_{(0,0)}$ is an infinitesimal rotation. Thus, any linear combination of these three vectors is ether an infinitesimal rotation or an infinitesimal translation. □

Exercise 3.4. *Fill in the details of this proof, by verifying that the above equation and the following equation both hold for all (x, y):*

$$a\rho_{(0,0)} + b\tau_{(1,0)} + c\tau_{(0,1)} = \begin{cases} \tau_{(b,c)}, & \text{if } a = 0; \\ a\rho_{(c/a,-b/a)}, & \text{if } a \neq 0. \end{cases}$$

The reader may find it interesting to verify that infinitesimal translations and rotations interact in the same way as the actual translations and rotations; see the following exercise.

Exercise 3.5. *Prove the following statements.*

1. *The sum of an infinitesimal translation and an infinitesimal rotation of magnitude a is another infinitesimal rotation of magnitude a (about a different center).*

2. *The sum of an infinitesimal rotation of magnitude a and an infinitesimal rotation of magnitude b (not necessarily with the same centers) is a rotation of magnitude $a+b$—unless $a+b=0$, in which case the sum is an infinitesimal translation.*

In our intuitive introduction of rigidity, we noted that the rigid motions of a planar framework were simply rigid motions of the plane restricted to the framework. The same is true of the infinitesimal rigid motions of a normal planar framework. Actually it is much easier to work with frameworks that are in general position—a stronger condition than simply being normal:

- A set $P = \{\mathbf{p}_1, \ldots, \mathbf{p}_n\}$ in the plane is in *general position* if no two points are equal and no three lie on a line.

In 1-space, general position simply means that no two points are equal; in 3-space, general position means no two points are equal, no three points lie on a line and no four points lie on a plane. Whenever the set of points $P = \{\mathbf{p}_1, \ldots, \mathbf{p}_{|V|}\}$ is in general position, we say that the framework $(V, E, \mathbf{p})$ is in *general position.* It should be clear that a general framework is a normal framework.

Lemma 3.6. *Let $\mathcal{F} = (V, E, \mathbf{p})$ be any planar framework in general position with $|V| \geq 2$. Then $\mathcal{R}(\mathcal{F})$ is the 3-dimensional space of infinitesimal translations and rotations restricted to the points in $P = \{\mathbf{p}_1, \ldots, \mathbf{p}_n\}$.*

Proof. Let $\mathbf{q} \in \mathcal{R}(\mathcal{F})$. Let $V = \{a_1, \ldots, a_n\}$ and denote the coordinates of $\mathbf{p}_i$ by (x_i, y_i) and the coordinates of $\mathbf{q}_i$ by (u_i, v_i), for all i. Clearly, the restriction of the translation $\tau_{(u_1,v_1)}$ to P is a rigid motion of $\mathcal{F}$. Hence $\hat{\mathbf{q}} = \mathbf{q} - \tau_{(u_1,v_1)} \in \mathcal{R}(\mathcal{F})$. Denote the coordinates of $\hat{\mathbf{q}}_i$ by $(\hat{u}_i, \hat{v}_i)$, for all i, and note that $\hat{\mathbf{q}}_1 = (0, 0)$. The condition

$$(x_2 - x_1, y_2 - y_1) \cdot (\hat{u}_2 - 0, \hat{v}_2 - 0) = 0$$

gives us that $\hat{\mathbf{q}}_2 = a(y_2 - y_1, x_1 - x_2)$, where $a = \hat{u}_2/(y_2 - y_1)$ when $y_1 \neq y_2$ and $a = \hat{v}_2/(x_1 - x_2)$ otherwise. (Since the points are in general position, either $y_1 \neq y_2$ or $x_1 \neq x_2$.)

Now define

$$\check{\mathbf{q}} = \hat{\mathbf{q}} - a\rho_{(x_1,y_1)} = \mathbf{q} - \tau_{(u_1,v_1)} - a\rho_{(x_1,y_1)};$$

as above, $\check{\mathbf{q}} \in \mathcal{R}(\mathcal{F})$. Furthermore, $\check{\mathbf{q}}_1 = (0, 0)$ and $\check{\mathbf{q}}_2 = (0, 0)$. Finally, for $i > 2$, we must have

$$(x_i - x_1, y_i - y_1) \cdot (\check{u}_i - 0, \check{v}_i - 0) = 0$$

and

$$(x_i - x_2, y_i - y_2) \cdot (\check{u}_i - 0, \check{v}_i - 0) = 0.$$

Since (x_i, y_i) is not on the line through (x_1, y_1) and (x_2, y_2), the vectors $(x_i - x_1, y_i - y_1)$ and $(x_i - x_2, y_i - y_2)$ are not parallel, whereas $(\check{u}_i, \check{v}_i)$ is perpendicular to both of them. Hence the only solution to the above equations is $(\check{u}_i, \check{v}_i) = (0, 0)$. We conclude that $\check{\mathbf{q}}$ assigns the zero vector to each point of $\mathcal{F}$ and, therefore, $\mathbf{q}$ is the restriction of $\tau_{(u_1,v_1)} + a\rho_{(x_1,y_1)}$ to P.

One minor thing remains to be proved, namely, that the space of infinitesimal translations and rotations restricted to the points in P is still 3-dimensional; in other words the function that maps every infinitesimal rigid motion of the plane to its restriction to P is a one-to-one linear transformation (we have just shown it to be onto). We leave that verification as an exercise for the reader. □

Exercise 3.6. *For any set of points P in the plane, let $\mathcal{R}(P)$ denote the space of infinitesimal translations and rotations restricted to the points in P.*

1. *Show that, when $|P| = 1$, $\mathcal{R}(P)$ is a 2-dimensional space.*
2. *Show that, when $|P| > 1$, $\mathcal{R}(P)$ is a 3-dimensional space.*

Exercise 3.7. *By an infinitesimal rigid motion of the plane, we mean a transformation $\sigma : R^2 \to R^2$ so that, for any pair of points (x, y) and $(\hat{x}, \hat{y})$, $(\hat{x} - x, \hat{y} - y) \cdot \big(\sigma(\hat{x}, \hat{y}) - \sigma(x, y)\big) = 0$. Adapt the proof of Lemma 3.6 to show that the only infinitesimal rigid motions of the plane are the infinitesimal translations and rotations.*

Since $\mathcal{R}(\mathcal{F})$ is a 3-dimensional subset of $\mathcal{M}(\mathcal{F})$, we will have $\mathcal{M}(\mathcal{F}) = \mathcal{R}(\mathcal{F})$ if and only if $\mathcal{M}(\mathcal{F})$ is also 3-dimensional. Thus:

Theorem 3.7. *Let $\mathcal{F} = (V, E, \mathbf{p})$ be any planar framework in general position with $|V| \geq 2$. Then $\dim[\mathcal{M}(\mathcal{F})] \geq 3$, and $\mathcal{F}$ is infinitesimally rigid if and only if $\dim[\mathcal{M}(\mathcal{F})] = 3$*

Recall the example of $(V, E, \mathbf{p})$, the rectangle in Figure 3.10. We showed that $\mathcal{R}(V, E, \mathbf{p})$ was the 3-dimensional subspace of vectors of the form $(\frac{2}{3}b + c - \frac{2}{3}d, d, \frac{2}{3}b + c - \frac{2}{3}d, b, c, b, c, d)$, for arbitrary $b, c, d \in R$. We may now select and interpret a basis:

- $(1, 0, 1, 0, 1, 0, 1, 0)$, $\tau_{(1,0)}$ restricted to $P = \{\mathbf{p}_1, \mathbf{p}_2, \mathbf{p}_3, \mathbf{p}_4\}$
  take $c = 1$, $b = d = 0$);
- $(0, 1, 0, 1, 0, 1, 0, 1)$, $\tau_{(0,1)}$ restricted to P
  (take $c = 0$, $b = d = 1$);
- $(0, 0, 0, 3, -2, 3, -2, 0)$, $\rho_{(0,0)}$ restricted to P
  (take $b = 3$, $c = -2$, $d = 0$).

In this example, $\mathcal{M}(V, E, \mathbf{p})$, consisting of all vectors of the form (a, d, a, b, c, b, c, d), for arbitrary $a, b, c, d \in R$, has dimension four. So we can extend the above basis for $\mathcal{R}(V, E, \mathbf{p})$ to a basis for $\mathcal{M}(V, E, \mathbf{p})$ by adding

- $(0, 0, 0, 0, 1, 0, 1, 0)$, the infinitesimal shear to the right, (take $c = 1$, $a = b = d = 0$).

This last vector is an infinitesimal motion but not an infinitesimal rigid motion. If we return to the infinitesimal deformation pictured in Figure 3.10, we may now write that deformation in terms of this basis:

$$\begin{aligned}\tfrac{1}{2}(1, -1, 1, 1, 1, 1, 1, 1, -1)\\ &= \tfrac{1}{2}(1, 0, 1, 0, 1, 0, 1, 0) - \tfrac{1}{2}(0, 1, 0, 1, 0, 1, 0, 1)\\ &\quad + \tfrac{1}{3}(0, 0, 0, 3, -2, 3, -2, 0) - \tfrac{2}{3}(0, 0, 0, 0, 1, 0, 1, 0).\end{aligned}$$

Exercise 3.8. *Consider Exercise 3.1 and write each motion of the framework in terms of the above basis. Note that, for each rigid motion, the shear is not used.*

We may now build a sound mathematical foundation for the "degrees of freedom" approach to rigidity. Consider a framework $\mathcal{F} = (V, E, \mathbf{p})$ in general position. We say that $\mathcal{F}$ has

- $\dim[\mathcal{M}(\mathcal{F})]$ degrees of freedom

and

- $\dim[\mathcal{M}(\mathcal{F})] - \dim[\mathcal{R}(\mathcal{F})]$ internal degrees of freedom.

By tying degrees of freedom to infinitesimal rigidity instead of rigidity, we avoid the problem with rigidity illustrated by the example in Figure 3.8, that is, frameworks held rigid by tension. If degrees of freedom were tied to rigidity, adding a single edge could reduce the count by more than one. But adding a single edge to a framework adds a single linear equation to the list of linear equations defining $\mathcal{M}(\mathcal{F})$ and can reduce the dimension of $\mathcal{M}(\mathcal{F})$ by at most one.

The concept of *implied edge* is also easy to understand in this context: An edge is implied by the framework $\mathcal{F}$ if the linear equation associated with the edge is a linear combination of the linear equations associated with the other edges of $\mathcal{F}$.

We employ this degrees of freedom analysis to prove the following corollary to Theorem 3.7.

Corollary 3.8. *Let $(V, E, \mathbf{p})$ be any planar framework with $|V| \geq 2$ and in general position. If $|E| < 2|V| - 3$, then the framework is not infinitesimally rigid.*

Proof. Let $\mathcal{F} = (V, E, \mathbf{p})$ be a planar framework in general position where $E = \{e_1, \ldots, e_{|E|}\}$ and let $E_i = \{e_1, e_2, \ldots, e_i\}$. Starting with

$$\mathcal{M}(V, E_0, \mathbf{p}) = R^{2|V|},$$

the entire space, and including the segments one at a time, each segment of the framework *may* reduce the dimension by at most one. Thus,

$$\dim[\mathcal{M}(V, E_i, \mathbf{p})] \geq 2n - i$$

and

$$\dim[\mathcal{M}(V, E, \mathbf{p})] \geq 2|V| - |E|.$$

We conclude that $\mathcal{F}$ can be infinitesimally rigid only if $3 \geq 2|V| - |E|$, that is, only if $|E| \geq 2|V| - 3$. □

We have worked through a detailed development of infinitesimal rigidity in dimensions one and two. What about 3-dimensional infinitesimal rigidity? In view of Lemma 3.2, the critical step in developing 3-dimensional infinitesimal rigidity is the computation of the dimension of $\mathcal{R}(V, E, \mathbf{p})$ for a 3-dimensional framework $(V, E, \mathbf{p})$. At the very end of Section 1.4, you were asked to supply a rationale for the conclusion that a rigid body in 3-space had 6 degrees of freedom; that is, that $\dim[\mathcal{R}(\mathcal{F})] = 6$ for any general framework containing three noncollinear points. This can now be demonstrated using the techniques developed for dimension two, proving the following analogues to Theorem 3.7 and its corollary:

Theorem 3.9. *Let $\mathcal{F} = (V, E, \mathbf{p})$ be any spatial framework in general position with $|V| \geq 3$. Then $\dim[\mathcal{M}(\mathcal{F})] \geq 6$, and $\mathcal{F}$ is infinitesimally rigid if and only if $\dim[\mathcal{M}(\mathcal{F})] = 6$,*

Corollary 3.10. *Let $(V, E, \mathbf{p})$ be any spatial framework with $|V| \geq 3$ and in general position. If $|E| < 3|V| - 6$, then the framework is not infinitesimally rigid.*

The interested reader should find it instructive to prove these results by adapting the proofs for dimension two. The proofs are outlined in the next exercise.

Exercise 3.9.

1. *Define the infinitesimal translations of space and show that they form a 3-dimensional vector space by identifying the three unit infinitesimal translations in the coordinate directions as a basis.*

2. *Define the unit infinitesimal rotations about each of the three axes and show that these three infinitesimal rotations along with the three basis vectors for the space of infinitesimal translations are independent and, therefore, form the basis of a 6-dimensional space of infinitesimal rigid motions of 3-space.*

   [A geometric note: Unlike the 2-dimensional case, this space of infinitesimal translations and rotations contains infinitesimal rigid motions that are neither infinitesimal translations nor infinitesimal rotations. Indeed the most common infinitesimal rigid motion in 3-space is the initial velocities of a screw-like motion, the simultaneous rotation about an axis and translation along that axis.]

3. *Prove the following Lemma 3.11, which is the 3-dimensional analogue to Lemma 3.6.*

4. *Use that lemma to prove Theorem 3.9 and its corollary.*

Lemma 3.11. *Let $\mathcal{F} = (V, E, \mathbf{p})$ be any spatial framework in general position with $|V| \geq 3$. Then $\mathcal{R}(\mathcal{F})$ is the 6-dimensional space of infinitesimal translations and rotations restricted to the points in $P = \{\mathbf{p}_1, \ldots, \mathbf{p}_n\}$.*

Exercise 3.10. *Consider the cubes with diagonal braces discussed in Section 1.1, pictured in Figure 1.2 and defined as frameworks in Exercise 2.4.*

1. *Show that the cube with four diagonal braces (pictured on the right in the figure) is not infinitesimally rigid. Keep the bottom square fixed and apply an infinitesimal rotation to the top square.*

2. *Show that the cube with all faces braced (pictured in the center of the figure) is infinitesimally rigid. Simplify your work by assuming that one of the triangles of the base is fixed, and verify that the only infinitesimal motion that leaves this triangle fixed leaves the entire cube fixed.*

We close this section with a sketch of a unified approach to infinitesimal rigidity, one that does not distinguish between normal and constricted frameworks in the definitions. The details are left as exercises for the interested reader. Before giving the new definition we must extend our original approach a bit. The first exercise gives the precise relationship between normal and general frameworks.

Exercise 3.11. *Prove that an m-dimensional framework $\mathcal{F} = (V, E, \mathbf{p})$ is normal if and only if $|V| \leq m$ and $\mathcal{F}$ is general or $|V| > m$ and $\mathcal{F}$ contains $m + 1$ points in general position.*

Next we must adapt the proof of Lemma 3.6 to prove the stronger result:

Exercise 3.12. *Let $\mathcal{F} = (V, E, \mathbf{p})$ be any normal planar framework with $|V| \geq 2$. Then $\mathcal{R}(\mathcal{F})$ is the 3-dimensional space of infinitesimal translations and rotations restricted to the points in $P = \{\mathbf{p}_1, \ldots, \mathbf{p}_n\}$.*

Our new approach starts with a complete development of the infinitesimal rigid motions of m-space. Then the infinitesimal rigid motions of an m-dimensional framework are defined to be the restrictions of the infinitesimal rigid motions of m-space to the points of the framework. For example, if $m = 2$, one must prove Exercise 3.7 and then take the statement of Exercise 3.12 as a definition. Having adopted this definition, several things still remain to be proven.

Exercise 3.13. *Let $\mathcal{F} = (V, E, \mathbf{p})$ be any normal planar framework with $|V| \geq 2$. Using the above definition of $\mathcal{R}(\mathcal{F})$, prove that $\mathcal{R}(\mathcal{F})$ is the space of all infinitesimal motions $\mathbf{q}$ of $\mathcal{F}$ such that $(\mathbf{p}_i - \mathbf{p}_j) \cdot (\mathbf{q}_i - \mathbf{q}_j) = 0$, for all $a_i, a_j \in V$.*

Exercise 3.14. *Let $\mathcal{F} = (V, E, \mathbf{p})$ be any constricted planar framework. Using the above definition of $\mathcal{R}(\mathcal{F})$, prove that $\mathcal{F}$ is not infinitesimally rigid.*

Once these two results have been proved, all proofs based on the old definition of 2-dimensional infinitesimal rigidity are valid.

Exercise 3.15. *Prove the* 1- *and* 3-*dimensional versions of these last two exercises.*

3.3 The Combinatorial Aspects of Rigidity

As we have seen (Theorems 2.16 and 3.3), rigidity and infinitesimal rigidity agree for all 1-dimensional frameworks in general position and are entirely combinatorial. That is, one can decide if a 1-dimensional framework in general position is rigid or infinitesimally rigid by simply checking its structure graph—specifically, checking whether its structure graph is connected or not.

Unfortunately, things are not so simple in higher dimensions. We have already shown that rigidity and infinitesimal rigidity do not always agree

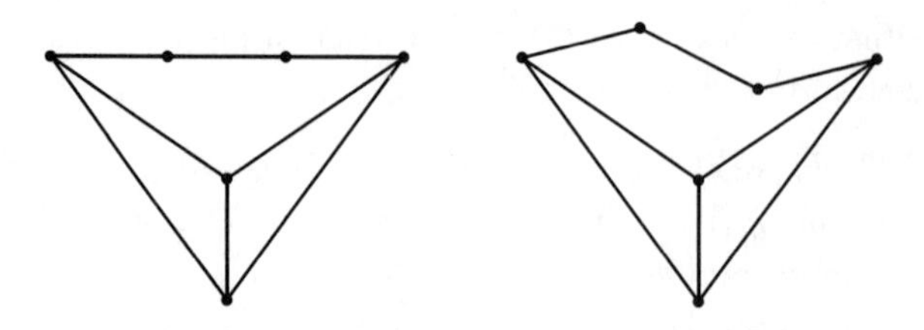

FIGURE 3.12

in dimension two, and we will soon show that neither of these concepts is combinatorial for 2-dimensional frameworks. On the other hand, we will also see that, in some sense, they are "almost" combinatorial. We start our discussion of the combinatorial aspects of rigidity in dimension two by reviewing an earlier example.

The framework on the left in Figure 3.12 is rigid (but not infinitesimally rigid) because of the very special positioning of its points. In the second picture of this figure, we have another framework with the same structure graph that is clearly not rigid: The "chain" of segments is now loose and can be twisted around. In fact, any framework in general position with this structure graph is not rigid. We might say that "almost all" frameworks with this structure graph are not rigid.

Now consider two other frameworks with identical structure graphs, as illustrated in Figure 3.13. To understand the motions of the left-hand framework, temporarily remove the edge joining the point labeled (x, y) and the point labeled $(\hat{x}, \hat{y})$. Now the framework consists of a rectangle (vertices labeled q, r, s, t) with two triangles attached. We note that the only deformations of this framework are obtained by deforming the rectangle. Without loss of generality, we may fix the bottom triangle and move only the top. As this framework moves, the segment with endpoints (x, y) and r and the segment

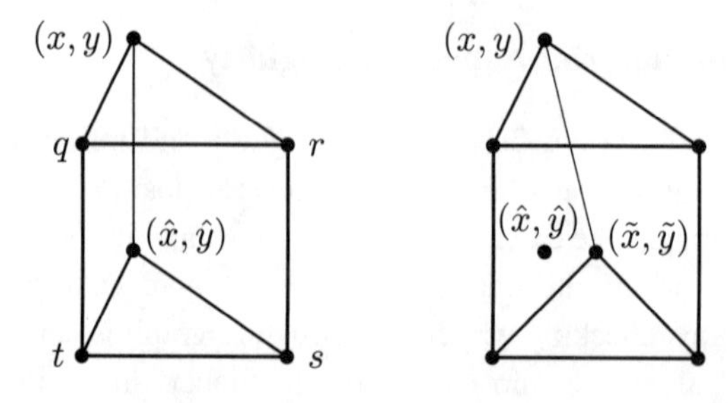

FIGURE 3.13

with endpoints $(\hat{x}, \hat{y})$ and s must remain parallel. Hence, (x, y), r, s and $(\hat{x}, \hat{y})$ will always be the vertices of a parallelogram; the edge joining (x, y) and $(\hat{x}, \hat{y})$ could have been left in, and the left-hand framework is not rigid.

Turning to the right-hand framework, again temporarily remove the edge joining the points labeled (x, y) and $(\tilde{x}, \tilde{y})$. Note that, when fixing the bottom triangle, the motions of the top triangle are the same as in the framework on the left. Hence (x, y) rotates in a circle centered at the point $(\hat{x}, \hat{y})$. Therefore the distance between (x, y) and $(\tilde{x}, \tilde{y})$ will change throughout the motion, and retaining the edge joining (x, y) and $(\tilde{x}, \tilde{y})$ will prevent the motion. So, the right-hand framework is rigid.

By Theorem 3.1, we see that the left-hand framework is also not infinitesimally rigid. To see that the right-hand framework is infinitesimally rigid in addition to being rigid requires a separate proof. Let $\mathbf{q}$ be any infinitesimal motion of the right-hand framework $\mathcal{F}$. Note that the restriction of $\mathbf{q}$ to the vertices of the bottom triangle must be an infinitesimal rigid motion. Since any infinitesimal rigid motion is the restriction of an infinitesimal rigid motion of the plane, there is an infinitesimal rigid motion $\mathbf{r}$ of $\mathcal{F}$ that agrees with $\mathbf{q}$ on the bottom triangle. Thus $\hat{\mathbf{q}} = \mathbf{q} - \mathbf{r}$ is also an infinitesimal motion of $\mathcal{F}$, and it assigns the zero vector to each vertex of the bottom triangle. Since $\hat{\mathbf{q}}$ assigns the zero vector to t, the vector it assigns to q must be perpendicular to the segment joining t and q. In fact, it must assign the same horizontal vector to q and r. That assignment can be extended to (x, y) in only one way: The same horizontal vector must be assigned to (x, y) also. Since $\hat{\mathbf{q}}$ assigns the zero vector to $(\hat{x}, \hat{y})$, the only horizontal vector that it can assign to (x, y) that will not stretch or compress the segment joining (x, y) and $(\hat{x}, \hat{y})$ is the zero vector. We conclude that $\hat{\mathbf{q}}$ assigns the zero vector to all the points of the framework and that $\mathbf{q} = \mathbf{r}$ is an infinitesimal rigid motion. We conclude that all of the infinitesimal motions of the framework are infinitesimal rigid motions, and the framework is infinitesimally rigid.

Consider the structure graphs of the frameworks we have just discussed. They are pictured in Figure 3.14. Looking at either one of them, what can we say about the class of planar frameworks with that graph as structure graph?

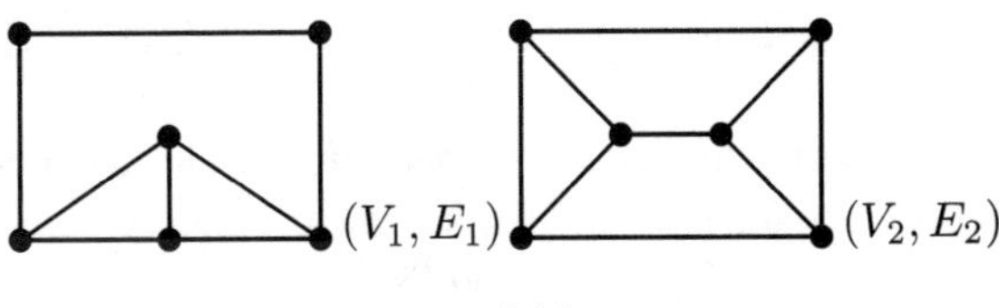

FIGURE 3.14

Consider a framework $(V_2, E_2, \mathbf{p})$ with structure graph (V_2, E_2). It seems that, unless the embedding $\mathbf{p}$ is very special, this framework will be infinitesimally rigid and therefore rigid. However, for very special embeddings, it will not be infinitesimally rigid nor be rigid. For a framework $(V_1, E_1, \mathbf{p})$ with (V_1, E_1) as structure graph, we can say a bit more: None of these frameworks are infinitesimally rigid. By the Corollary to Theorem 3.7, (V_1, E_1) does not have enough edges to be infinitesimally rigid, as

$$|E| = 8 < 9 = 2|V| - 3.$$

However, if the embedding $\mathbf{p}$ is very special, the framework $(V_1, E_1, \mathbf{p})$ could be rigid.

These two examples illustrate a very fundamental fact: If the embedding $\mathbf{p}$ of a framework $(V, E, \mathbf{p})$ is not special, then either the framework is both rigid and infinitesimally rigid, or the framework is neither rigid nor infinitesimally rigid. Which of these two possibilities holds depends entirely on the structure graph (V, E).

In short, for nonspecial embeddings, the concepts of rigidity and infinitesimal rigidity agree and are combinatorial!

The embeddings that are not special with respect to rigidity or infinitesimal rigidity are called *generic* embeddings. Understanding generic embeddings is the key to understanding the combinatorial aspects of rigidity. Since this is a rather complicated concept, we precede the development of generic embeddings with a discussion of the related but simpler concept of *general* embeddings. Recall that:

- a collection of points in the line is in general position if the points are distinct;
- a collection of points in the plane is in general position if the points are all distinct and no three lie on a straight line; and
- a collection of points in 3-space is said to be in general position if the points are all distinct, no three lie on a straight line and no four lie in a plane.

Lemma 3.12. *Almost all m-dimensional frameworks are in general position.*

The difficulty with this lemma is that "almost all" has not been defined; furthermore, we do not have the tools (measure theory) necessary to give a formal definition. However, we can build a picture of sets in general position that will justify this lemma. We start by considering the two simple examples

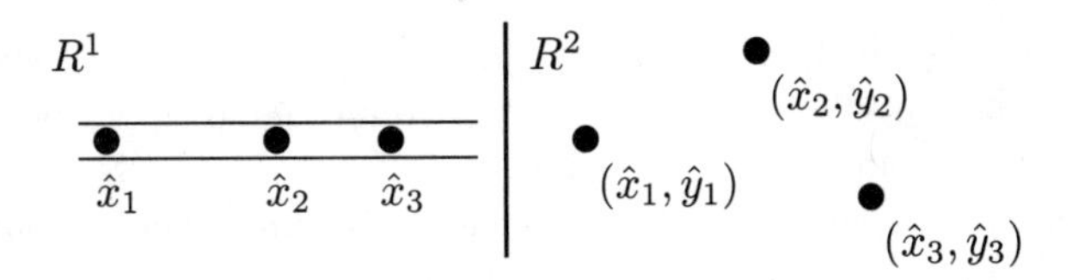

FIGURE 3.15

in Figure 3.15. On the left we have a general embedding of $V = \{a_1, a_2, a_3\}$ in the line and, on the right, a general embedding of V in the plane.

As we have discussed before, we may identify each embedding of V into R^m with a single point in $R^{m|V|}$. The left-hand embedding of the figure is identified with the point $(\hat{x}_1, \hat{x}_2, \hat{x}_3)$ in R^3 and the right-hand embedding of the figure is identified with the point $(\hat{x}_1, \hat{y}_1, \hat{x}_2, \hat{y}_2, \hat{x}_3, \hat{y}_3)$ in R^6. Concentrating on the first example, let (x_1, x_2, x_3) represent an arbitrary embedding of V in the line. Our question is: When will (x_1, x_2, x_3) represent an embedding that is *not* general?

On the line, the only way that an embedding can fail to be general is for one or more of the image points to coincide. That is, the points in R^3 corresponding to nongeneral embeddings lie on one of the planes given by the equations $x_1 = x_2$, $x_1 = x_3$ and $x_2 = x_3$. In Figure 3.16, we have sketched a portion of the plane, $x_1 = x_2$. A plane has volume zero, as does the union of three planes. Clearly then, "almost all" of the points in R^3 correspond to general embeddings, while only those relative few that lie in this set of volume zero correspond to nongeneral embeddings.

Turning to embeddings of V in the plane, we would like to compile a description of the points of R^6 that correspond to the nongeneral embeddings. First are the conditions that imply that two of the points coincide. For example,

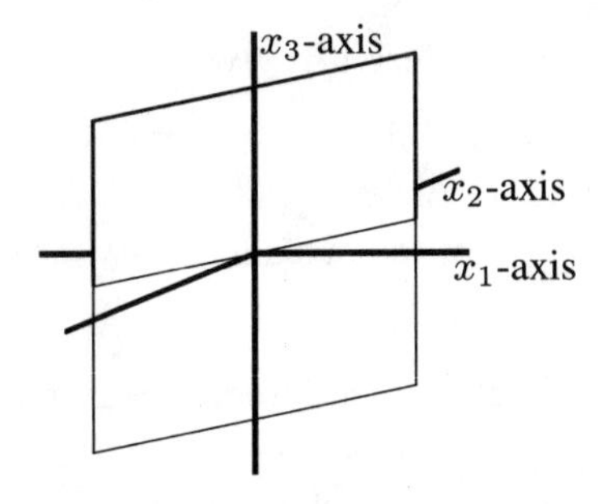

FIGURE 3.16

the solution set to the system $\{x_1 = x_2, y_1 = y_2\}$ is a 4-dimensional subspace of R^6. So the embeddings in which two or more points coincide lie on the union of three 4-dimensional subspaces.

The only other way an embedding of V in the plane can be nongeneral is if the three points are collinear. Recall that

$$\big(\lambda x_2 + (1-\lambda)x_1, \lambda y_2 + (1-\lambda)y_1\big)$$

is a parameterization of the line through (x_1, y_1) and (x_2, y_2). (We have that $\lambda = 0$ gives (x_1, y_1) and $\lambda = 1$ gives (x_2, y_2).) Thus, the three points will be collinear if and only if

$$x_3 = \lambda x_2 + (1-\lambda)x_1$$

and

$$y_3 - \lambda y_2 + (1-\lambda)y_1$$

for some value of λ.

Suppose that $x_1 \neq x_2$ and $y_1 \neq y_2$. Then solving each equation for λ gives

$$\frac{x_3 - x_1}{x_2 - x_1} = \lambda = \frac{y_3 - y_1}{y_2 - y_1}.$$

Eliminating λ, cross multiplying and reorganizing the terms, we get

$$x_1y_2 + x_2y_3 + x_3y_1 = x_1y_3 + x_2y_1 + x_3y_2.$$

Now if $x_1 = x_2$, then $x_1 = x_2 = x_3$ and this equation holds. Similarly, it holds when $y_1 = y_2$. The solutions to this equation form a 5-dimensional, quadratic surface in 6-space:

- The surface is quadratic since it is given by a quadratic equation.
- It is 5-dimensional since, with few exceptions, five of the six variables may be assigned values independently; and the point with those five coordinates and the sixth, computed to satisfy the equation, will then lie on the surface.

Thus, almost all embeddings of three vertices are general: all points in R^6 but those corresponding to points on the union of three 4-dimensional subspaces and a 5-dimensional, quadratic surface.

Now let $\mathbf{p}$ be any planar embedding of $V = \{a_1, \ldots, a_n\}$. For $\mathbf{p}$ to be a general embedding, the point in R^{2n} corresponding to $\mathbf{p}$ must avoid the $\binom{n}{2}$

$(2n-2)$-dimensional subspaces given by the systems

$$\begin{aligned} x_j - x_i &= 0, \\ y_j - y_i &= 0, \end{aligned} \qquad \text{for } 1 \leq i < j \leq n,$$

as well as the $\binom{n}{3}$ $(2n-1)$-dimensional quadratic surfaces given by the quadratic equations

$$x_i y_j + x_j y_k + x_k y_i = x_i y_k + x_j y_i + x_k y_j,$$

for $1 \leq i < j < k \leq n$.

Exercise 3.1. *Consider the spatial embeddings of V in R^3 where $V = \{a_1, \ldots, a_n\}$.*

1. *Let $n = 4$ and describe the sets of points in R^{12} that correspond to nongeneral embeddings.*
2. *For arbitrary $n \geq 4$, describe the sets of points in R^{3n} that correspond to nongeneral embeddings.*

As we have noted, generic embeddings are far more difficult to define. Intuitively, they are all of the embeddings that avoid the special positions where frameworks behave erratically. These special positions are of two kinds:

- those that misbehave relative to both rigidity and infinitesimal rigidity, as illustrated in Figure 3.13; and
- those that misbehave relative to rigidity alone, as illustrated in Figure 3.12.

It is not too difficult to describe a finite collection of subspaces and surfaces of lower dimensions that contains all embeddings that misbehave relative to both rigidity and infinitesimal rigidity. Indeed, we will do that at the end of this section. While it is true that all embeddings that misbehave relative to rigidity alone also lie on a finite collection of subspaces and surfaces of lower dimensions, describing such a collection requires mathematics well beyond the prerequisites for this text.

We will proceed with this intuitive definition of "generic": a generic set of points in R^m, a generic embedding and a generic framework. We assert:

> *All points of R^{mn} that correspond to the special or nongeneric embeddings of $V = \{a_1, \ldots, a_n\}$ into R^m ($m = 2, 3$) lie on a finite collection of subspaces and surfaces of lower dimensions.*

This fact, which we must accept without proof, implies the first part of the next theorem.

Theorem 3.13. *For dimensions one, two and three.*

1. *Almost all frameworks are generic.*
2. *Rigidity and infinitesimal rigidity agree on generic frameworks. That is, if $(V, E, \mathbf{p})$ is a generic framework, then it is either both rigid and infinitesimally rigid or neither rigid nor infinitesimally rigid.*
3. *Either all generic frameworks with the same structure graph are infinitesimally rigid and rigid, or all are neither infinitesimally rigid nor rigid.*

Statement (2) tells us that the differences between rigidity and infinitesimal rigidity are restricted to the relatively small collection of special (nongeneric) frameworks and that, for almost all frameworks, these concepts agree. Statement (3) tells us that, for generic frameworks, rigidity (equal to infinitesimal rigidity) is *combinatorial* and hence a property of the underlying structure graph.

Taken together, these three facts lead to the definition of rigidity for graphs:

> A graph (V, E) is said to be *rigid in dimension m*, or simply *m-rigid*, if some (and hence all) generic m-dimensional frameworks with it as structure graph are infinitesimally rigid and rigid.

None of the three statements of the theorem is obvious for dimensions greater than one. Perhaps Statement (3) is the most surprising: One may suspect that two generic frameworks with the same structure graphs, but embedded quite differently (like the cubic frameworks pictured in Figure 2.8) may actually behave differently contrary to Statement (3). As we have said, a complete proof of all three parts of this theorem requires techniques that are beyond the scope of this book. However, we can prove Statements (1) and (3) as they relate to infinitesimal rigidity alone. We close this section with an outline of such a proof, leaving some of the details to be filled in by the interested reader.

Let the m-dimensional framework $\mathcal{F} = (V, E, \mathbf{p})$ be given, and consider the space of its infinitesimal motions, $\mathcal{M}(\mathcal{F})$. As we have seen, this is the solution space of a homogeneous system of $|E|$ equations in a set of $m|V|$ variables, the coordinates of the vector assignment $\mathbf{q} : V \to R^m$.

Next, consider the matrix of coefficients of this system. Its entries are simple (linear) functions of the coordinates of the embedding $\mathbf{p} : V \to R^m$.

We call this $|E| \times m|V|$ matrix the *rigidity matrix* of the framework $\mathcal{F}$ and denote it by $\mathbf{M}(\mathcal{F})$.

Now let's consider a specific example. Let $m = 2$, let $V = \{a_1, a_2, a_3\}$, let $E = \{\{a_1, a_2\}, \{a_1, a_3\}\}$ and let (x_i, y_i) denote the coordinates of the points of the embedding $\mathbf{p} : V \to R^2$. If $\mathcal{F}$ denotes the framework $(V, E, \mathbf{p})$, we have the system of homogeneous equations

$$(x_1 - x_2)u_1 + (y_1 - y_2)v_1 + (x_2 - x_1)u_2 + (y_2 - y_1)v_2 = 0$$

$$(x_1 - x_3)u_1 + (y_1 - y_3)v_1 + (x_3 - x_1)u_3 + (y_3 - y_1)v_3 = 0$$

and the corresponding rigidity matrix

$$\mathbf{M}(\mathcal{F}) = \begin{bmatrix} (x_1 - x_2) & (y_1 - y_2) & (x_2 - x_1) & (y_2 - y_1) & 0 & 0 \\ (x_1 - x_3) & (y_1 - y_3) & 0 & 0 & (x_3 - x_1) & (y_3 - y_1) \end{bmatrix}.$$

For an arbitrary framework $\mathcal{F}$ in m-space, $\mathcal{M}(\mathcal{F})$ is the kernel or null space of the rigidity matrix, $\mathbf{M}(\mathcal{F})$. And by a basic result from linear algebra,

$$\dim\left[\mathcal{M}(\mathcal{F})\right] + \operatorname{rank}\left[\mathbf{M}(\mathcal{F})\right] = m|V|.$$

This observation permits us to restate Theorems 3.7 and 3.9 in terms of the rank of the rigidity matrix:

Corollary 3.14. [to Theorem 3.7] *Let $\mathcal{F} = (V, E, \mathbf{p})$ be any planar framework in general position with $|V| \geq 2$. Then* rank $\left(\mathbf{M}(\mathcal{F})\right) \leq 2|V| - 3$, *and $\mathcal{F}$ is infinitesimally rigid if and only if*

$$\operatorname{rank}\left(\mathbf{M}(\mathcal{F})\right) = 2|V| - 3.$$

Corollary 3.15. [to Theorem 3.9] *Let $\mathcal{F} = (V, E, \mathbf{p})$ be any spatial framework in general position with $|V| \geq 3$. Then* rank $\left(\mathbf{M}(\mathcal{F})\right) \leq 3|V| - 6$, *and $\mathcal{F}$ is infinitesimally rigid if and only if*

$$\operatorname{rank}\left(\mathbf{M}(\mathcal{F})\right) = 3|V| - 6.$$

Returning to the plane, we illustrate these results with the nongeneric and generic frameworks pictured in Figure 3.13. We redraw those frameworks below with labels indicating a specific embedding for each.

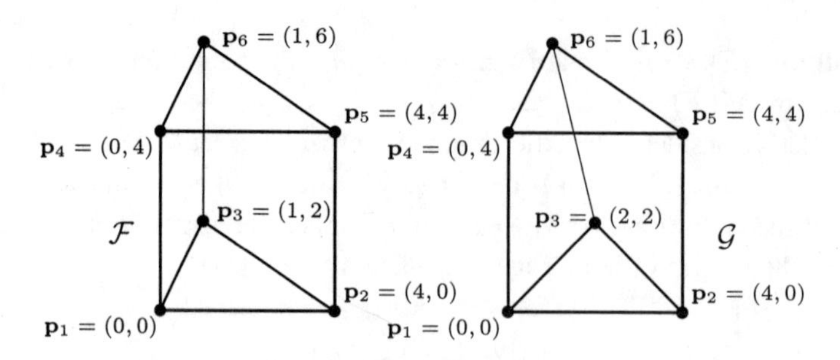

FIGURE 3.17

Exercise 3.2. *Consider the frameworks $\mathcal{F}$ and $\mathcal{G}$ as pictured in Figure 3.17 In each case let $\mathbf{q} : V \to R^2$ denote a vector assignment to the points of the framework. (To avoid cluttering the figure, the vectors $\mathbf{q}_i = (u_i, v_i)$ are not pictured.)*

1. *Verify that this is the rigidity matrix for $\mathcal{F}$, $\mathbf{M}(\mathcal{F})$:*

$$\left|\begin{array}{cccccccccccc} 0 & -4 & 0 & 0 & 0 & 0 & 0 & 4 & 0 & 0 & 0 & 0 \\ 0 & 0 & 0 & 0 & 0 & 0 & -4 & 0 & 4 & 0 & 0 & 0 \\ 0 & 0 & 0 & 0 & 0 & 0 & -1 & -2 & 0 & 0 & 1 & 2 \\ 0 & 0 & 0 & 0 & 0 & 0 & 0 & 0 & 3 & -2 & -3 & 2 \\ 0 & 0 & 0 & -4 & 0 & 0 & 0 & 0 & 0 & 4 & 0 & 0 \\ -4 & 0 & 4 & 0 & 0 & 0 & 0 & 0 & 0 & 0 & 0 & 0 \\ -1 & -2 & 0 & 0 & 1 & 2 & 0 & 0 & 0 & 0 & 0 & 0 \\ 0 & 0 & 3 & -2 & -3 & 2 & 0 & 0 & 0 & 0 & 0 & 0 \\ 0 & 0 & 0 & 0 & 0 & -4 & 0 & 0 & 0 & 0 & 0 & 4 \end{array}\right| \begin{array}{c} (\mathbf{p}_1 - \mathbf{p}_4) \\ (\mathbf{p}_4 - \mathbf{p}_5) \\ (\mathbf{p}_4 - \mathbf{p}_6) \\ (\mathbf{p}_5 - \mathbf{p}_6) \\ (\mathbf{p}_2 - \mathbf{p}_5) \\ (\mathbf{p}_1 - \mathbf{p}_2) \\ (\mathbf{p}_1 - \mathbf{p}_3) \\ (\mathbf{p}_2 - \mathbf{p}_3) \\ (\mathbf{p}_3 - \mathbf{p}_6) \end{array}$$

   [Note: The columns correspond to the variables $u_1, v_1, u_2, v_2, \ldots, u_6, v_6$ in that order; for each row, the corresponding edge of the framework is identified to the right of the matrix.]

2. *Verify that* rank $\big(\mathbf{M}(\mathcal{F})\big) = 8$.

3. *Compute* $\mathbf{M}(\mathcal{G})$. [Note that $\mathbf{M}(\mathcal{F})$ and $\mathbf{M}(\mathcal{G})$ differ only in the last three rows.]

4. *Verify that* rank $\big(\mathbf{M}(\mathcal{G})\big) = 9$.

In this example, the rank of the rigidity matrix of the nongeneric framework is smaller than it is for the generic embedding. This is the key to understanding generic positions relative to infinitesimal rigidity. Before giving a formal definition, we work through one more example.

Let $V = \{a_1, a_2, a_3, a_4\}$ and let (x_i, y_i) denote the coordinates of an embedding $\mathbf{p} : V \to R^2$. Let $\mathcal{F}$ denote the framework, $(V, K, \mathbf{p})$, with the complete graph (V, K) as structure graph. The rigidity matrix for $\mathbf{M}(\mathcal{F})$ is

$$\begin{bmatrix} (x_1-x_2) & (y_1-y_2) & (x_2-x_1) & (y_2-y_1) & 0 & 0 & 0 & 0 \\ (x_1-x_3) & (y_1-y_3) & 0 & 0 & (x_3-x_1) & (y_3-y_1) & 0 & 0 \\ 0 & 0 & (x_2-x_3) & (y_2-y_3) & (x_3-x_2) & (y_3-y_2) & 0 & 0 \\ (x_1-x_4) & (y_1-y_4) & 0 & 0 & 0 & 0 & (x_4-x_1) & (y_4-y_1) \\ 0 & 0 & (x_2-x_4) & (y_2-y_4) & 0 & 0 & (x_4-x_2) & (y_4-y_2) \\ 0 & 0 & 0 & 0 & (x_3-x_4) & (y_3-y_4) & (x_4-x_3) & (y_4-y_3) \end{bmatrix}$$

Consider the $k \times k$ minors of $\mathbf{M}(\mathcal{F})$ for all k between 1 and 6. Recall that a $k \times k$ minor of an $m \times n$ matrix $\mathbf{M}$ is the $k \times k$ matrix obtained by deleting any $m - k$ rows and any $n - k$ columns from $\mathbf{M}$. The minors of the rigidity matrix are of two types:

1. those minors that have a zero determinant only for special values of the variables; and

2. those with identically zero determinants.

For example, the 2×2 minor listed in Exercise 3.3 is of the first type while the 3×3 minor listed there is of the second type.

Exercise 3.3. *Verify the following.*

1. $$\det \begin{bmatrix} (x_1 - x_2) & (x_2 - x_1) \\ (x_1 - x_3) & 0 \end{bmatrix} = 0,$$

   if and only if $x_1^2 - x_1x_2 - x_1x_3 + x_2x_3 = 0$.

2. $$\det \begin{bmatrix} (x_1 - x_2) & (x_2 - x_1) & 0 \\ (x_1 - x_3) & 0 & (x_3 - x_1) \\ 0 & (x_2 - x_3) & (x_3 - x_2) \end{bmatrix} = 0,$$

   for all x_1, x_2, x_3.

Now, to each $k \times k$ minor of the first type, we associate the "surface" that is the solution to the polynomial equation of degree k obtained by setting its determinant equal to zero. If the point in R^8 corresponding to $\mathbf{p}$ avoids all of these surfaces, we say that the embedding $\mathbf{p}$ or the collection of points $\{\mathbf{p}_1, \mathbf{p}_2, \mathbf{p}_3, \mathbf{p}_4,\}$ is *generic*.

To see that this definition corresponds to our intuitive notion of generic, consider any framework $\mathcal{G} = (V, E, \mathbf{p})$ on this set of points. Note that $\mathbf{M}(\mathcal{G})$ consists of those rows of $\mathbf{M}(\mathcal{F})$ that correspond to the edges in E. Note also that the rank of $\mathbf{M}(\mathcal{G})$ is the size of its largest minor with a nonzero determinant. Finally, note that this rank remains the same for all generic $\mathbf{p}$ and can only be smaller for nongeneric $\mathbf{p}$.

We now give the formal definition of generic. Let $m = 2$ or 3; let $V = \{a_1, \ldots, a_n\}$ with $n > m$; let $\mathbf{p} : V \to R^m$; let the coordinates of $\mathbf{p}_i$ given by (x_i, y_i) or (x_i, y_i, z_i); let (V, K) be the complete graph on V; and let $\mathbf{M}$ denote the rigidity matrix of $(V, K, \mathbf{p})$. Let $\mathcal{S}$ denote the union of the surfaces in R^{mn} that are the solution sets to the polynomial equations obtained by setting to zero the determinants of the minors of $\mathbf{M}$ that are not identically zero. Under these conditions:

- An embedding $\mathbf{p}$ whose corresponding point in R^{mn} does not lie on $\mathcal{S}$ is said to be *generic.*
- All frameworks with a generic embedding are *generic frameworks.*

It is important to note that it is the *embedding* of a generic framework that is generic. Hence any other framework with the same embedding (or the same set of points) must also be generic. Also we should note that, if $\mathbf{p} : V \to R^m$ is generic and $V' \subseteq V$, then $\mathbf{p}|_{V'}$ is also generic. This follows at once from the fact that the rigidity matrix of the "complete" framework $(V', K', \mathbf{p}|_{V'})$ is a submatrix of the rigidity matrix of $(V, K, \mathbf{p})$.

Another very useful consequence of this definition of generic is given by:

Theorem 3.16. *Let $m = 2$ or 3 and let $(V, E, \mathbf{p})$ be any m-dimensional framework (not necessarily generic). If $(V, E, \mathbf{p})$ is infinitesimally rigid, then any generic framework in m-space with (V, E) as structure graph is infinitesimally rigid and the graph (V, E) is m-rigid.*

Furthermore, if the graph (V, E) is not m-rigid, then any m-dimensional framework $(V, E, \mathbf{p})$ with it as structure graph is not infinitesimally rigid.

To simplify the proof of this theorem, note that the expressions

$$(|V| - 1), \quad \text{in dimension one,}$$

$$(2|V| - 3), \quad \text{in dimension two, and}$$

$$(3|V| - 6), \quad \text{in dimension three,}$$

can be restated succinctly as

$$m|V| - \binom{m+1}{2}, \quad \text{in dimension } m, \quad \text{for } m = 1, 2, 3.$$

Proof. Let $(V, E, \mathbf{p})$ be any m-dimensional infinitesimally rigid framework. By Theorems 3.7 and 3.9,

$$\text{rank } \big(\mathbf{M}(V, E, \mathbf{p})\big) = m|V| - \binom{m+1}{2}.$$

Now let $(V, E, \hat{\mathbf{p}})$ be any generic, m-dimensional framework with the same structure graph. By the definition of generic,

$$\text{rank } \big(\mathbf{M}(V, E, \hat{\mathbf{p}})\big) \geq \text{ rank } \big(\mathbf{M}(V, E, \mathbf{p})\big)$$

and, by Theorems 3.7 and 3.9,

$$m|V| - \binom{m+1}{2} \geq \text{ rank } \big(\mathbf{M}(V, E, \hat{\mathbf{p}})\big).$$

It follows that

$$\text{rank } \big(\mathbf{M}(V, E, \hat{\mathbf{p}})\big) = m|V| - \binom{m+1}{2}.$$

Thus $(V, E, \hat{\mathbf{p}})$ is infinitesimally rigid, and the graph (V, E) is rigid in dimension m.

Observe that the second statement of the theorem is simply the contrapositive of the first. □

Let's stand back and look at what we have accomplished: We have developed a purely combinatorial concept, m-dimensional rigidity, for graphs. Furthermore, if a graph is not rigid in dimension m, then no m-dimensional framework based on it can be infinitesimally rigid and only in very exceptional cases could such a framework be rigid. In the early planning stages of a structure, exact dimensions are seldom fixed. Hence, we are actually working with the structure graph. If we wish the final framework to be rigid and infinitesimally rigid, we will certainly want the structure graph to be rigid!

However, at this point in our exposition, we still have no combinatorial methods for working with m-dimensional rigidity for graphs when $m > 1$. At present, the only way we have of checking whether a graph is m-rigid, for $m > 1$, is to construct a generic framework R^m with that graph as structure graph and then check whether this framework is infinitesimally rigid. In the next section of this chapter, we will develop two combinatorial methods for checking directly whether or not a graph is generically 2-rigid.

We close this section with a useful observation, stated in the following lemma.

Lemma 3.17. *Every generic embedding is a general embedding.*

The proof of this lemma is left as an exercise for the interested reader; the major steps in the proof are outlined below.

Exercise 3.4. *Let* $\mathbf{p}$ *be an embedding of* V *into* m*-space* ($m = 1, 2, 3$)*. Let* $a_0, a_1, a_2, a_3 \in V$ *and let* $\mathbf{p}_i = (x_i)$ *or* $\mathbf{p}_i = (x_i, y_i)$ *or* $\mathbf{p}_i = (x_i, y_i, z_i)$*. Verify each of the following steps:*

1. *If* $\mathbf{p}$ *is generic, the determinant of the* 1×1 *minor* $[x_0 - x_1]$ *is not zero.*

2. *If* $\mathbf{p}_0 = \mathbf{p}_1$*, the determinant of this minor is zero.*

3. *If* $\mathbf{p}$ *is generic and* $m = 2$ *or* 3*, the determinant of the* 2×2 *minor*

$$\begin{bmatrix} (x_0 - x_1) & (y_0 - y_1) \\ (x_0 - x_2) & (y_0 - y_2) \end{bmatrix}$$

is not zero.

4. *If* $\mathbf{p}_0, \mathbf{p}_1$ *and* $\mathbf{p}_2$ *are collinear, the determinant of this minor is zero.*

5. *If* $\mathbf{p}$ *is generic and* $m = 3$*, the determinant of the* 3×3 *minor*

$$\begin{bmatrix} (x_0 - x_1) & (y_0 - y_1) & (z_0 - z_1) \\ (x_0 - x_2) & (y_0 - y_2) & (z_0 - z_2) \\ (x_0 - x_3) & (y_0 - y_3) & (z_0 - z_3) \end{bmatrix}$$

is not zero.

6. *If* $\mathbf{p}_0, \mathbf{p}_1, \mathbf{p}_2$ *and* $\mathbf{p}_3$ *are coplanar, the determinant of this minor is zero.*

3.4 Rigidity for Graphs

By its definition, rigidity for graphs is still tied to infinitesimal rigidity. If we want to check whether a graph (V, E) is m-rigid ($m = 2$ or 3), we must select a generic embedding $\mathbf{p} : V \to R^m$ and check whether the framework $(V, E, \mathbf{p})$ is infinitesimally rigid. Clearly, it would be to our advantage to be able to characterize rigidity for graphs in terms of some easily checked properties of the graph. We have done this in the case of 1-dimensional rigidity: A graph (V, E) is 1-rigid if and only if it is connected.

Being connected can be easily checked without constructing a generic framework in the line with the given graph as structure graph. In fact, several characterization of connectivity exist. Although most of these characterizations do not have nice generalizations to higher dimensions, one approach to connectivity generalizes well: *Connected graphs have been characterized in terms of trees.* Specifically:

- A graph is connected if and only if it has a spanning tree. (Lemma 2.7)

Furthermore, trees have been characterized in three ways:

- A graph (V, E) is a tree if and only if (V, E) is connected and $|E| = |V| - 1$. (Lemma 2.10)
- A graph (V, E) is a tree if and only if the following conditions hold:
    1. $|E(U)| \leq |U| - 1$, for all $\emptyset \subset U \subseteq V$.
    2. $|E| = |V| - 1$. (Theorem 2.19)
- A graph (V, E) is a tree if and only if it can be constructed from a single vertex by a sequence of 1-extensions. (Theorem 2.20)

We start by generalizing the concept of a tree to higher dimensions. For $m = 1$, 2 or 3, a graph (V, E) is an *m-tree* if it is m-rigid but the removal of any edge results in a graph that is not m-rigid. Thus (V, E) is an m-tree if and only if it is the structure graph of an m-dimensional, generic, isostatic framework. If (V, E) is m-rigid but not an m-tree, it must have an edge e so that $(V, E - e)$ is also m-rigid. If $(V, E - e)$ is not an m-tree, we may delete another edge e', resulting in the m-rigid, spanning subgraph $(V, E - e - e')$. This process must eventually stop with a spanning m-tree.

We have proved the generalization of Lemma 2.7:

Lemma 3.18. *For $m = 1$, 2 or 3, a graph is m-rigid if and only if it has an m-tree as a spanning subgraph.*

Next we state and prove the generalization of Lemma 2.10 to m-trees.

Lemma 3.19. *For $m = 1$, 2 or 3, a graph $(V; E)$ that satisfies any two of the following conditions satisfies the third and is an m-tree; furthermore, any m-tree satisfies these three conditions.*

1. *(V, E) is m-rigid.*
2. *The rows of the rigidity matrix of any generic embedding of V into m-space are independent.*
3. $|E| = m|V| - \binom{m+1}{2}$.

Proof. Let $\mathbf{p} : V \to R^m$ be any generic embedding of V into m-space. By Corollaries 3.14 and 3.15, we may restate condition (1) as

1. rank $\big(\mathbf{M}(V, E, \mathbf{p})\big) = m|V| - \binom{m+1}{2}$.

Next, we note that $\mathbf{M}(V, E, \mathbf{p})$ has $|E|$ rows. Thus condition (2) may be restated as

2. rank $\big(\mathbf{M}(V, E, \mathbf{p})\big) = |E|$.

It is now obvious that, if any two of conditions (1), (2) and (3) hold, so does the third.

Next, suppose (V, E) satisfies any two and hence all three of these conditions. By (1), (V, E) is m-rigid. Now let e be any edge in E. By condition (3), we have

$$\text{rank}\,\big(\mathbf{M}(V, E - e, \mathbf{p})\big) \leq |E| - 1 < |E| = m|V| - \binom{m+1}{2};$$

so $(V, E - e)$ is not m-rigid. We conclude that (V, E) is an m-tree.

Finally, let (V, E) be an m-tree. Condition (1) clearly holds. Suppose condition (2) did not, and the row of $\mathbf{M}(V, E, \mathbf{p})$ corresponding to the edge $e \in E$ was a linear combination of the remaining rows. Then

$$\text{rank}\,\big(\mathbf{M}(V, E - e, \mathbf{p})\big) = \text{rank}\,\big(\mathbf{M}(V, E, \mathbf{p})\big),$$

and $(V, E - e)$ is also m-rigid, contradicting the fact that (V, E) is an m-tree. So condition (2) must also hold. □

Exercise 3.1. *To complete the analogy between Lemmas 3.19 and 2.10, show that condition* (2) *of Lemma 3.19 is equivalent to* (V, E) *being circuit-free when* $m = 1$.

Next we consider a generalization to higher dimensions of half of Theorem 2.19.

Lemma 3.20. [The Laman Conditions] *Let* (V, E) *be an* m*-tree with* $|V| \geq m$*, for* $m = 1$*,* 2 *or* 3*. Then*

1. $|E(U)| \leq m|U| - \binom{m+1}{2}$*, for all* $U \subseteq V$ *with* $|U| \geq m$*, and*
2. $|E| = m|V| - \binom{m+1}{2}$.

Proof. This result has already been proved in the case $m = 1$. So let $m = 2, 3$ and let (V, E) be an m-tree, with $|V| \geq m$. The second condition follows from the previous lemma. Let $\mathbf{p} : V \to R^m$ be a generic embedding of V and assume $U \subseteq V$ and $|U| \geq m$. Then, by Corollary 3.14 and Corollary 3.15, the matrix $\mathbf{M}\big(U, E(U), \mathbf{p}\big)$ has rank at most $m|U| - \binom{m+1}{2}$. Since (V, E) is an m-tree, the rows of $\mathbf{M}(V, E, \mathbf{p})$, and hence the rows of $\mathbf{M}\big(V, E(U), \mathbf{p}\big)$ are independent. But the rows of $\mathbf{M}\big(U, E(U), \mathbf{p}\big)$ are simply the rows of $\mathbf{M}(V, E(U), \mathbf{p})$ with some columns of zeros deleted. We conclude that the rows of $\mathbf{M}\big(U, E(U), \mathbf{p}\big)$ are independent and

$$E(U) = \text{ rank } \Big(\mathbf{M}\big(U, E(U), \mathbf{p}\big)\Big) \leq m|U| - \tbinom{m+1}{2}. \ \square$$

The full 2-dimensional generalization of Theorem 2.19 follows.

Theorem 3.21. [Laman, 1975] *A graph (V, E) is a 2-tree if and only if*

1. $|E(U)| \leq 2|U| - 3$, *for all $U \subseteq V$ with $|U| \geq 2$, and*
2. $|E| = 2|V| - 3$.

It is because of this important result that the conditions in Lemma 3.20 are called the Laman Conditions. We will not prove Theorem 3.21 just yet. Instead, we move on to a consideration of the m-dimensional generalization of Theorem 2.20. Let $m = 1$, 2 or 3, let (V, E) be a graph, with $V = \{a_1, \ldots, a_n\}$, $n \geq m$, let a_0 be a vertex distinct from the vertices in V and let $\{a_{i_1}, \ldots, a_{i_m}\}$ be any collection of m vertices. Then the graph $(\hat{V}, \hat{E})$, where $\hat{V} = \{a_0, a_1, \ldots, a_n\}$, and $\hat{E} = E \cup \big\{\{a_0, a_{i_1}\}, \ldots, \{a_0, a_{i_m}\}\big\}$ is called an *m-extension* of (V, E).

Lemma 3.22. *Any m-extension of an m-tree is an m-tree. Conversely, deleting any vertex of valence m from an m-tree results in a smaller m-tree.*

Proof. Let $m = 1$, 2 or 3, let (V, E) be a graph, with $V = \{a_1, \ldots, a_n\}$, and let $(\hat{V}, \hat{E})$ with $\hat{V} = \{a_0, a_1, \ldots, a_n\}$ and $\hat{E} = E \cup \big\{\{a_0, a_{i_1}\}, \ldots, \{a_0, a_{i_m}\}\big\}$ be an m-extension of (V, E). Finally, let $\mathbf{p}$ be a generic embedding of $\hat{V}$ into m-space and denote its restriction to V by $\mathbf{p}|_V$. As we have already noted, $\mathbf{p}|_V$ is also generic.

Suppose first that (V, E) is m-rigid and let $\mathbf{q}$ be any infinitesimal motion of $(\hat{V}, \hat{E})$. Since (V, E) is m-rigid, the restriction of $\mathbf{q}$ to V is an infinitesimal rigid motion of $(V, E, \mathbf{p}|_V)$. But then, by Lemmas 3.6 and 3.11, the restriction of $\mathbf{q}$ to V is an infinitesimal rigid motion of m-space restricted to V. Let $\mathbf{r}$

denote that infinitesimal rigid motion of m-space restricted to $\hat{V}$. Then $\mathbf{q}-\mathbf{r}$ is an infinitesimal motion of $(\hat{V}, \hat{E}, \mathbf{p})$. Since $\mathbf{q}-\mathbf{r}$ assigns the zero vector to every vertex in V, $\mathbf{q}_0 - \mathbf{r}_0$ must be orthogonal to each of the m vectors $(\mathbf{p}_{i_1} - \mathbf{p}_0), \ldots, (\mathbf{p}_{i_m} - \mathbf{p}_0)$. Since the embedding is generic, the points $\mathbf{p}_0$, $\mathbf{p}_{i_1}, \ldots, \mathbf{p}_{i_m}$ are in general position and these vectors are independent. We conclude that $\mathbf{q}_0 - \mathbf{r}_0$ must be the zero vector. Thus $\mathbf{q} = \mathbf{r}$ and $\mathbf{q}$ is an infinitesimal rigid motion of $(\hat{V}, \hat{E}, \mathbf{p})$. We conclude that if (V, E) is m-rigid then $(\hat{V}, \hat{E})$ is m-rigid.

Conversely, assume that $(\hat{V}, \hat{E})$ is m-rigid. Thus $\mathbf{M}(\hat{V}, \hat{E}, \mathbf{p})$ has rank

$$m|\hat{V}| - \binom{m+1}{2} = m|V| - \binom{m+1}{2} + m.$$

Now $\mathbf{M}(V, E, \mathbf{p})$ is obtained from $\mathbf{M}(\hat{V}, \hat{E}, \mathbf{p})$ by first deleting the m rows corresponding to the m edges containing a_0, then deleting the m columns of zeros corresponding to the coordinates of $\mathbf{p}_0$. Hence the rank of $\mathbf{M}(V, E, \mathbf{p}|_V)$ is at least

$$m|V| - \binom{m+1}{2}$$

and (V, E) is m-rigid.

So, we have shown that (V, E) is m-rigid if and only if $(\hat{V}, \hat{E})$ is m-rigid. Since $|\hat{E}| = |E| + m$ and $m|\hat{V}| = m|V| + m$, we see that

$$|E| = m|V| - \binom{m+1}{2}$$

if and only if

$$|\hat{E}| = m|\hat{V}| - \binom{m+1}{2}.$$

Thus by Lemma 3.19 (V, E) is an m-tree if and only if $(\hat{V}, \hat{E})$ is an m-tree. □

Next we consider a construction called "edge splitting." Let $m = 1, 2$ or 3, let (V, E) be a graph, with $V = \{a_1, \ldots, a_n\}$, $n > m$, let a_0 be a vertex distinct from the vertices in V and let $\{a_{i_1}, a_{i_2}, \ldots, a_{i_{m+1}}\}$ be any $m+1$ vertices with $\{a_{i_1}, a_{i_2}\} = e \in E$. Then the graph $(\hat{V}, \hat{E})$, where $\hat{V} = \{a_0, a_1, \ldots, a_n\}$, and $\hat{E} = (E - e) \cup \{\{a_0, a_{i_1}\}, \ldots, \{a_0, a_{i_{m+1}}\}\}$ is called an m-dimensional edge split or an *m-edge-split* of (V, E). A 1-edge-split for the graph (V, E) is simply the insertion of a new vertex at the midpoint of one of its edges; a 2-edge-split is the insertion of a new vertex at the midpoint of one of its edges followed by including an edge between the new vertex and an third vertex of the graph. A 3-edge-split is pictured in the left-hand diagram in Figure 3.18.

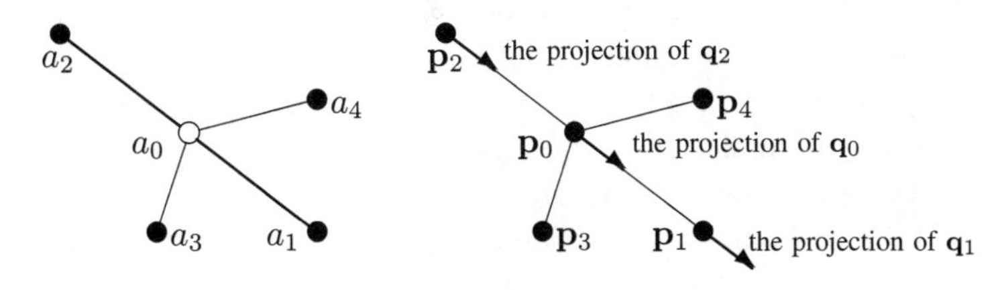

FIGURE 3.18

Lemma 3.23. *Any m-edge-split of an m-tree is an m-tree. Conversely, if one deletes a vertex a_0 of valence $m+1$ from an m-tree, then one may add an edge between one of the pairs of vertices adjacent to a_0 so that the resulting graph will be an m-tree.*

Proof. Let $m = 1$, 2 or 3, let (V, E) be a graph and let $(\hat{V}, \hat{E})$ be an m-edge-split of (V, E). We adopt the above notation with the simplifying assumption that a_0 is attached to $a_1, \ldots, a_{m+1}$.

Suppose first that (V, E) is m-rigid and let $\mathbf{p}$ be any generic embedding of V into m-space. Since $\mathbf{p}$ is a generic embedding, $\{\mathbf{p}_1, \mathbf{p}_2, \ldots, \mathbf{p}_n\}$ is in general position. In the exercise following this proof, you will show that we can find a point $\mathbf{p}_0$ interior to the segment joining $\mathbf{p}_1$ and $\mathbf{p}_2$ so that $\{\mathbf{p}_0, \mathbf{p}_2, \ldots, \mathbf{p}_n\}$ is in general position. Assume that $\mathbf{p}_0$ has been selected in this manner, and extend $\mathbf{p}$ to an embedding of $\hat{V}$ into m-space by mapping a_0 onto $\mathbf{p}_0$ (see the right-hand diagram in Figure 3.18 for the case $m = 3$). Note that this new embedding is no longer generic; but, as we will see, it serves our purposes very well. Let $\mathbf{q}$ be an infinitesimal motion of $(\hat{V}, \hat{E}, \mathbf{p})$.

We may argue geometrically that, since $\mathbf{p}_1$, $\mathbf{p}_0$ and $\mathbf{p}_2$ are collinear, the projections of $\mathbf{q}_1$, $\mathbf{q}_0$ and $\mathbf{q}_2$ onto the vectors $\mathbf{p}_2 - \mathbf{p}_0$ and $\mathbf{p}_0 - \mathbf{p}_1$ must all be equal. Arguing algebraically, we have $(\mathbf{p}_2 - \mathbf{p}_1) = \alpha_2(\mathbf{p}_2 - \mathbf{p}_0) = \alpha_1(\mathbf{p}_0 - \mathbf{p}_1)$. So

$$(\mathbf{p}_2 - \mathbf{p}_0) \cdot (\mathbf{q}_2 - \mathbf{q}_0) = 0 \quad \text{and} \quad (\mathbf{p}_0 - \mathbf{p}_1) \cdot (\mathbf{q}_0 - \mathbf{q}_1) = 0$$

imply

$$(\mathbf{p}_2 - \mathbf{p}_1) \cdot (\mathbf{q}_2 - \mathbf{q}_0) = 0 \quad \text{and} \quad (\mathbf{p}_2 - \mathbf{p}_1) \cdot (\mathbf{q}_0 - \mathbf{q}_1) = 0.$$

Adding these two equalities gives

$$(\mathbf{p}_2 - \mathbf{p}_1) \cdot (\mathbf{q}_2 - \mathbf{q}_1) = 0.$$

Using the geometric or the algebraic argument, we conclude that the restriction of $\mathbf{q}$ to V is an infinitesimal motion of $(V, E, \mathbf{p})$. But since (V, E) is m-rigid, the restriction of $\mathbf{q}$ to V is an infinitesimal rigid motion of $(V, E, \mathbf{p})$ and, therefore, an infinitesimal rigid motion of m-space restricted to V.

Let $\mathbf{r}$ denote that infinitesimal rigid motion of m-space restricted to $\hat{V}$. Then $\mathbf{q} - \mathbf{r}$ is an infinitesimal motion of $(\hat{V}, \hat{E}, \mathbf{p})$. Since $\mathbf{q} - \mathbf{r}$ assigns the zero vector to every vertex in V, $\mathbf{q}_0 - \mathbf{r}_0$ must be orthogonal to each of the m segments joining $\mathbf{p}_0$ to $\mathbf{p}_2, \ldots, \mathbf{p}_{m+1}$. Since $\mathbf{p}_0, \mathbf{p}_2, \ldots, \mathbf{p}_{m+1}$ is in general position, the direction vectors of the segments joining $\mathbf{p}_0$ to $\mathbf{p}_2, \ldots, \mathbf{p}_{m+1}$ are independent; we conclude that $\mathbf{q}_0 - \mathbf{r}_0$ must be the zero vector. Thus $\mathbf{q} = \mathbf{r}$ and $\mathbf{q}$ is an infinitesimal rigid motion of $(\hat{V}, \hat{E}, \mathbf{p})$. We conclude that, if (V, E) is m-rigid, then $(\hat{V}, \hat{E}, \mathbf{p})$ is infinitesimally rigid. It follows from Theorem 3.16 that $(\hat{V}, \hat{E})$ is also m-rigid.

Now suppose that (V, E) was actually an m-tree. Then

$$
\begin{aligned}
|\hat{E}| &= |E| + m \\
&= m|V| - \binom{m+1}{2} + m \\
&= m|\hat{V}| - \binom{m+1}{2}
\end{aligned}
$$

and $(\hat{V}, \hat{E})$ is an m-tree.

Conversely, assume that $(\hat{V}, \hat{E})$, is an m-tree with a vertex a_0 of valence $m+1$. Specifically, let $\hat{V} = \{a_0, a_1, \ldots, a_n\}$, let $V = \{a_1, \ldots, a_n\}$, let $E = E(V)$ and let $\hat{E} = E \cup \big\{\{a_0, a_1\}, \ldots, \{a_0, a_{m+1}\}\big\}$. Now let $\mathbf{p}$ be a generic embedding of $\hat{V}$. We must show that after deleting a_0 and the $m+1$ edges containing it, we can add an edge between one of the pairs among $a_1, \ldots, a_{m+1}$ to get an m-tree. We have that $\mathbf{M}(\hat{V}, \hat{E}, \mathbf{p})$ has rank

$$
m|\hat{V}| - \binom{m+1}{2} = m|V| - \binom{m+1}{2} + m.
$$

The matrix $\mathbf{M}(V, E, \mathbf{p})$ is obtained from $\mathbf{M}(\hat{V}, \hat{E}, \mathbf{p})$ by deleting the $m+1$ rows corresponding to the $m+1$ edges containing a_0 and then deleting the columns of zeros corresponding to a_0. Thus, rank $\big(\mathbf{M}(V, E, \mathbf{p})\big)$ is equal to the number of its rows or

$$
m|V| - \binom{m+1}{2} - 1.
$$

Since (V, E) does not have enough edges, it is not infinitesimally rigid, and it admits an infinitesimal deformation $\mathbf{q}$.

Our plan is to show that $\mathbf{q}$ must distort the distance between one of the pairs among $a_1, \ldots a_{m+1}$—say the pair $\{a_1, a_2\}$. From this we could conclude that the equation corresponding to the edge $\{a_1, a_2\}$ is independent of the equations corresponding to the edges of (V, E). Adding the row corresponding to this equation to $\mathbf{M}(V, E, \mathbf{p})$ gives $\mathbf{M}(V, E', \mathbf{p})$ for the graph (V, E'), where $E' = E \cup \{a_1, a_2\}$. Furthermore, it would follow that rank

$$(\mathbf{M}(V, E', \mathbf{p})) = m|V| - \binom{m+1}{2}$$

and, since $|E'| = m|V| - \binom{m+1}{2}$, that (V, E') is rigid.

To carry out this plan, we suppose that $\mathbf{q}$ does not distort any of the distances between pairs from among $a_1, \ldots a_{m+1}$ and we show that this supposition leads to a contradiction. Since $\mathbf{q}$ does not distort any of the distances between pairs from among $a_1, \ldots a_{m+1}$, its restriction to these points agrees with an infinitesimal rigid motion of m-space $\mathbf{r}$ on this set of points.

Now extend $\mathbf{q}$ to $\hat{V}$ by setting $\mathbf{q}_0 = \mathbf{r}_0$. Since the extended $\mathbf{q}$ distorts none of the additional edges $\{a_0, a_1\}, \ldots, \{a_0, a_{m+1}\}$, it is an infinitesimal motion of $(\hat{V}, \hat{E})$. But since it is an infinitesimal deformation of the subgraph (V, E), it is an infinitesimal deformation of $(\hat{V}, \hat{E})$, contradicting the fact that $(\hat{V}, \hat{E})$ is infinitesimally rigid. □

Exercise 3.2. *For $m = 1, 2, 3$, prove that, if the collection $\mathbf{p}_1, \mathbf{p}_2, \ldots, \mathbf{p}_{m+1}$ is in general position in m-space, then we may select a point $\mathbf{p}_0$ interior to the segment joining $\mathbf{p}_1$ and $\mathbf{p}_2$ so that the collection $\mathbf{p}_0, \mathbf{p}_2, \ldots, \mathbf{p}_{m+1}$ is also in general position.* [Hint: Show that the set of points $\mathbf{p}_0$ on the segment so that the collection $\mathbf{p}_0, \mathbf{p}_2, \ldots, \mathbf{p}_{m+1}$ is not in general position is finite.]

We are now ready to state and prove the 2-dimensional extension of Theorem 2.20.

Theorem 3.24. [Henneberg 1911] *A graph (V, E) is a 2-tree if and only if it may be constructed from a single edge by a sequence of 2-extensions and 2-edge-splits.*

Proof. By Lemmas 3.22 and 3.23, any graph constructed by a 2-extension or a 2-edge-split from a 2-tree is another 2-tree; thus any graph constructed from a single edge by a sequence of 2-extensions and 2-edge-splits is a 2-tree. To prove the converse, we proceed by induction on $|V|$. Clearly the only 2-tree on three vertices is constructed from a single edge by a 2-extension.

Let n be at least 4, assume that every 2-tree on fewer than n vertices may be constructed from a single edge by a sequence of 2-extensions and 2-edge-splits and let (V, E) be a 2-tree on n vertices. By Lemma 3.19, $|E| = 2|V| - 3$; then, by Lemma 2.1, the average vertex valence is

$$\frac{2|E|}{|V|} = 4 - \frac{6}{|V|} < 4.$$

It is also a simple consequence of the Laman Conditions that every vertex of a 2-tree on three or more vertices has valence at least 2. (See part 1 of Exercise 3.3 below.) We conclude that (V, E) has a vertex of valence 2 or a vertex of valence 3. If it has a vertex of valence 2, it is a 2-extension of a 2-tree (V', E'), by Lemma 3.22. If it has a vertex of valence 3, it is a 2-edge-split of a 2-tree (V', E'), by Lemma 3.23. Applying the induction hypothesis to (V', E') completes the proof. □

Exercise 3.3.

1. *Let* (V, E) *be a graph on three or more vertices that satisfies the* 2-*dimensional Laman Conditions, and verify the following statements:*
   - *Every vertex of* (V, E) *has valence at least* 2.
   - *At least one vertex of* (V, E) *has valence less than* 4.
2. *Let* (V, E) *be a graph on four or more vertices that satisfies the* 3-*dimensional Laman Conditions, and verify the following statements:*
   - *Every vertex of* (V, E) *has valence at least* 3.
   - *At least one vertex of* (V, E) *has valence less than* 6.

We can now use Henneberg's result, which we have just proved, to complete the proof of Laman's Theorem. (This is the longest proof in the book; it involves a variety of combinatorial and algebraic computations.)

Proof of Laman's Theorem. We have already proved that if (V, E) is a 2-tree then it must satisfy Laman's conditions for dimension two. We proceed by induction to prove that any graph that satisfies the 2-dimensional Laman Conditions is a 2-tree. This is easily checked to be so when $|V| \leq 3$. Assume then that $n \geq 3$ and that all graphs on n or fewer vertices satisfying the 2-dimensional Laman Conditions are 2-trees. Let (V, E), with $|V| = n + 1$, satisfy the 2-dimensional Laman Conditions. As we have argued before—and as you proved in Exercise 3.3—(V, E) contains a vertex of valence 2 or a vertex of valence 3.

Let $V = \{a_0, a_1, \ldots, a_n\}$, and assume that a_0 is the vertex of valence 2 or 3. Assume first that $\rho(a_0) = 2$, and assume that $\{a_0, a_1\}, \{a_0, a_2\} \in E$. Let $V' = V - a_0$ and $E' = E - \{a_0, a_1\}, \{a_0, a_2\}$. Then for $U \subseteq V'$ with $|U| \geq 2$, we have $U \subseteq V$; so

$$|E'(U)| = |E(U)| \leq 2|U| - 3.$$

Also,

$$|E'| = |E| - 2 = 2|V| - 5 = 2|V'| - 3.$$

We conclude that (V', E') satisfies the Laman Condition and, by the induction hypothesis, is a 2-tree. Clearly, (V, E) is a 2-extension of (V', E'), so it is also a 2-tree.

Next assume that $\rho(a_0) = 3$ and $\{a_0, a_1\}, \{a_0, a_2\}, \{a_0, a_3\} \in E$. Let $V' = V - \{a_0\}$ and $E' = E - \{\{a_0, a_1\}, \{a_0, a_2\}, \{a_0, a_3\}\}$. If we can show that one edge, $\{a_1, a_2\}$, $\{a_1, a_3\}$ or $\{a_2, a_3\}$, can be added to (V', E') without violating the Laman Conditions, then by the same chain of reasoning as above:

- (V', E') plus the edge will satisfy the Laman Conditions.
- By the induction hypothesis, it will be a 2-tree.
- As a 3-edge-split of this graph, (V, E) will also be a 2-tree.

So, the remainder of this proof is devoted to proving that one of the edges $\{a_1, a_2\}$, $\{a_1, a_3\}$ or $\{a_2, a_3\}$ can be added to (V', E') without violating the Laman Conditions. Specifically, we must show that, for one of the pairs $\{a_1, a_2\}$, $\{a_1, a_3\}$ or $\{a_2, a_3\}$, we have that $|E'(U)| < 2|U| - 3$ for all $U \subset V'$ containing that pair.

Suppose there is a set $U \subset V'$ such that $a_1, a_2, a_3 \in U$ and $|E'(U)| = 2|U| - 3$. But then

$$\begin{aligned}|E(U \cup \{a_0\})| &= |E'(U)| + 3 = (2|U| - 3) + 3 \\ &= 2|U \cup \{a_0\}| - 2 > 2|U \cup \{a_0\}| - 3,\end{aligned}$$

which is impossible.

Thus, for any $U \subset V'$ containing a_1, a_2 and a_3, we have

$$|E'(U)| < 2|U| - 3.$$

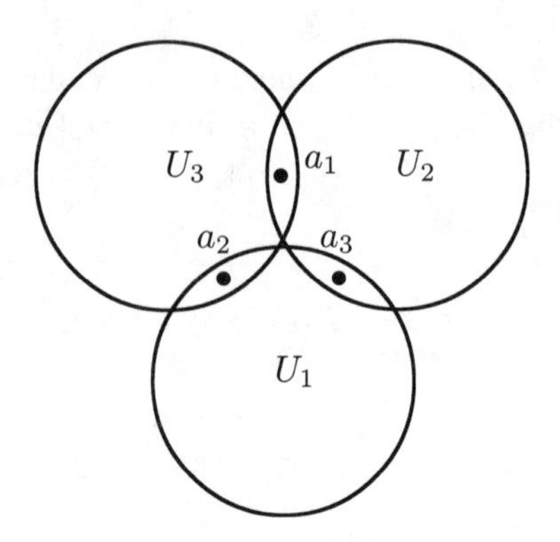

FIGURE 3.19

Now, suppose that there is a set $U_1 \subset V' - \{a_1\}$ containing a_2 and a_3 with $E'(U_1) = 2|U_1| - 3$ and, similarly, sets U_2 and U_3 with $E'(U_i) = 2|U_i| - 3$, for $i = 2, 3$. See Figure 3.19. We must show that this supposition leads to a contradiction. Our first step will be to show that $|U_i \cap U_j| = 1$ for all distinct i and j. Let $U = U_1 \cup U_2$. We assert

$$|E'(U)| \geq |E'(U_1)| + |E'(U_2)| - |E'(U_1 \cap U_2)|. \tag{1}$$

To prove (1), observe that $|E(U_1)|$ ($|E(U_2)|$) counts the edges with both endpoints in U_1 (U_2) and $-|E'(U_1 \cap U_2)|$ subtracts those edges with both endpoints in $U_1 \cap U_2$ and therefore is counted twice. Uncounted, and the reason for the inequality, are any edges with one endpoint in U_1, but not U_2, and the other in U_2, but not U_1.

Assuming $|U_1 \cap U_2| > 1$, we have

$$|E'(U_1 \cap U_2)| \leq 2|U_1 \cap U_2| - 3,$$

so

$$\begin{aligned} |E'(U)| &\geq |E'(U_1)| + |E'(U_2)| - |E'(U_1 \cap U_2)| \\ &\geq (2|U_1| - 3) + (2|U_2| - 3) - (2|U_1 \cap U_2)| - 3) \\ &= 2|U| - 3. \end{aligned}$$

But U contains a_1, a_2 and a_3, forcing $|E'(U)| < 2|U| - 3$, by our earlier argument. Hence, the assumption $|U_1 \cap U_2| > 1$ must be false, and $(U_1 \cap U_2) = \{a_3\}$. Similarly, $(U_i \cap U_j) = \{a_k\}$, for all $\{i, j, k\} = \{1, 2, 3\}$.

Finally, let $U = U_1 \cup U_2 \cup U_3$. Then

$$|U| = |U_1| + |U_2| + |U_3| - 3$$

and we have

$$\begin{aligned} |E'(U)| &\geq |E'(U_1)| + |E'(U_2)| + |E'(U_3)| \\ &\geq 2|U_1| - 3 + 2|U_2| - 3 + 2|U_3| - 3 \\ &= 2|U| - 3. \end{aligned}$$

But again U contains a_1, a_2 and a_3, forcing $|E(U)| < 2|U| - 3$. Hence, the original assumption that three such sets exist must be false. In other words, there must be one pair $\{a_i, a_j\}$ $(1 \leq i < j \leq 3)$ so that, for any $U \subset V'$ containing a_i and a_j, $|E'(U)| < 2|U| - 3$. $\square$

With Theorem 3.21 and Theorem 3.24, we may investigate generic 2-dimensional rigidity without the necessity of finding generic embeddings for our graphs. The next section is devoted to such an investigation.

3.5 Generic Rigidity in Dimension Two

With either of the Laman or Henneberg results, we can check small graphs directly for rigidity. Let's consider a few examples. First, the complete bipartite graph $K_{3,3}$: A graph on six vertices needs $2 \times 6 - 3 = 9$ edges to be a 2-tree; so $K_{3,3} = (V, E)$ is a candidate. Let's check it out using Laman's Theorem.

- As we have just noted, condition (2) of the theorem is satisfied.
- We must now verify that no subgraph on two vertices has more than one edge, that is,

  $$|E(U)| \leq 2 \times 2 - 3 = 1, \qquad \text{for } |U| = 2,$$

  and that no subgraph on three vertices has more than three edges,

  $$|E(U)| \leq 2 \times 3 - 3 = 3, \qquad \text{for } |U| = 3.$$

  But these conditions hold for every graph.

- Next we must show that no subgraph on four vertices has more than five edges,
$$|E(U)| \leq 2 \times 4 - 3 = 5, \qquad \text{for } |U| = 4.$$

  But a subgraph on four vertices with more than five edges is a copy of K_4, and $K_{3,3}$ contains no triangles much less a K_4.

- Finally we easily check that each subgraph on five vertices has six edges, one less than the maximum of $7 = 2 \times 5 - 3$ that is allowed.

So, by Laman's Theorem, $K_{3,3}$ is a 2-tree and any generic 2-dimensional framework with $K_{3,3}$ as structure graph will be rigid.

It is instructive to verify again that $K_{3,3}$ is a 2-tree, this time using Henneberg's constructive technique. Working backwards, one easily finds the Henneberg sequence for $K_{3,3}$. Such a sequence is pictured in Figure 3.20. It consists of three 2-extensions followed by one 2-edge-split. At each stage the new vertex is indicated by an open circle.

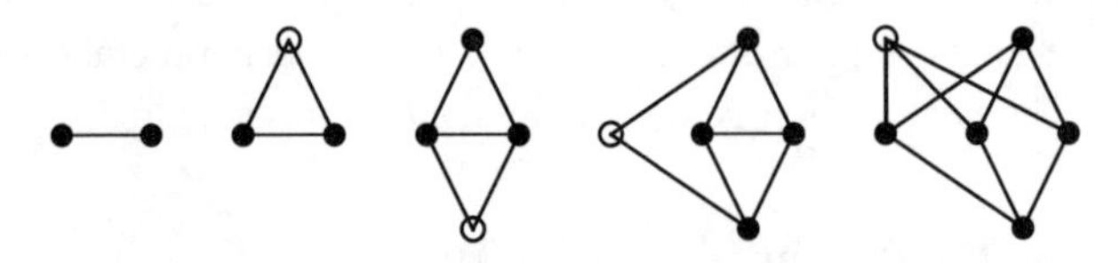

FIGURE 3.20

Recall that in Figure 1.11, we pictured all three graphs that can be produced by a 2-extension from the triangle or two 2-extensions from a single edge, that is, all 2-trees on five vertices that can be constructed by 2-extensions alone. Since the average valence of a 2-tree on five vertices is $\frac{2 \times 7}{5} < 3$, every 2-tree on five vertices can be constructed by 2-extensions alone. Figure 1.11 pictures all 2-trees on five vertices.

In Exercise 1.1, you were challenged to produce all eleven 2-trees on six vertices that can be constructed by 2-extensions alone. We have just shown that $K_{3,3}$ is also a 2-trees on six vertices. There is just one other 2-tree on six vertices that has no vertex of valence 2.

Exercise 3.1. *List all thirteen 2-trees on six vertices.*

The next two exercises include several other interesting examples.

Exercise 3.2. *For each of the following graphs, decide whether it is 2-rigid or not and whether it is a 2-tree or not.*

1. $K_{2,n}$ *plus any edge (two cases).*
2. $K_{3,4}$.
3. $K_{3,5}$ *minus any 2 edges (two cases).*

Exercise 3.3. *For each graph in Figure 3.21, decide whether it is 2-rigid or not and whether it is a 2-tree or not.*

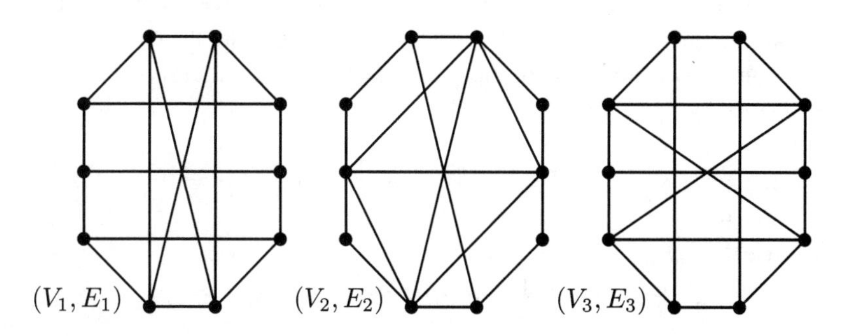

FIGURE 3.21

Another advantage in having a combinatorial characterization of 2-rigidity for graphs is that it makes proving results about 2-rigidity for graphs much easier. We illustrate this by proving several useful properties of 2-trees.

Lemma 3.25. *Let (V, E) be a 2-tree with $|V| \geq 3$. Then:*

1. *Every vertex of (V, E) has valence at least 2.*
2. *(V, E) is connected ((V, E) is 1-rigid).*
3. *Every edge of (V, E) lies on a circuit ((V, E) is 1-birigid).*

Proof. You were asked to prove the first statement in Exercise 3.3. We include it here for the sake of completeness. Let $a \in V$ and let $U = V - \{a\}$. Then $|E(U)| = |E| - \rho(a)$. By Laman's Theorem,

$$|E| - \rho(a) = |E(U)| \leq 2|U| - 3 = 2|V| - 2 - 3 = |E| - 2.$$

So $\rho(a) \geq 2$.

Next suppose that $(V_1, E_1), \ldots, (V_k, E_k)$ are the components of (V, E). In view of the previous part, each contains at least three vertices. Therefore,

$|E_i| \le 2|V_i| - 3$, for each i, and

$$|E| = |E_1| + \cdots + |E_k| \le 2(|V_1| + \cdots + |V_k|) - 3k = |V| - 3k.$$

But since $|E| = 2|V| - 3$, we conclude that $k = 1$.

Finally, let $e \in E$ and let $E' = E - \{e\}$. If we can show that (V, E') is connected, then we can conclude that the endpoints of e are joined by a path in (V, E') completing a circuit with e. To show this, we simply adapt the previous argument. Suppose that $(V_1, E_1), \ldots, (V_k, E_k)$ are the component of (V, E'). In view of first part of this proof, each contains at least two vertices. So, $|E_i| \le 2|V_i| - 3$, for each i, and

$$|E'| = |E_1| + \cdots + |E_k| \le 2(|V_1| + \cdots + |V_k|) - 3k = |V| - 3k.$$

But since

$$|E'| = |E| - 1 = (2|V| - 3) - 1 = 2|V| - 4,$$

we again conclude that $k = 1$. □

Exercise 3.4. *Formulate and prove a result stating that, with certain exceptions, removing two edges from a 2-tree cannot disconnect it.*

The next exercise lists a few simple results about gluing 2-trees together to get larger 2-trees. The reader is invited to prove them using Theorem 3.21 or Theorem 3.24.

Exercise 3.5. *Let (V, E) and (U, F) be two 2-trees, such that $V \cap U = \{a_1, a_2\}$. Prove the following.*

1. *If $\{a_1, a_2\}$ is an edge of both of these graphs, then $(V \cup U, E \cup F)$ is a 2-tree.*

2. *If $\{a_1, a_2\}$ is an edge of exactly one of the graphs, then $(V \cup U, E \cup F - \{a_1, a_2\})$ is a 2-tree.*

3. *If $\{a_1, a_2\}$ is an edge of neither of these graphs, then $(V \cup U, E \cup F)$ is 2-rigid but not a 2-tree.*

We say that a graph (V, E) is a 2-*circuit* if, for each edge $e \in E$, $(V, E - e)$ is a 2-tree. These 2-circuits have a nice Laman-like characterization, stated in the next theorem.

Theorem 3.26. *A graph (V, E) is a 2-circuit if and only if*

1. $|E(U)| \leq 2|U| - 3$, *for all* $U \subset V$ *with* $2 \leq |U| < |V|$, *while*
2. $|E| = 2|V| - 2$.

Proof. Suppose that (V, E) satisfies the above conditions and that e is any edge in E. It follows at once that $(V, E - e)$ satisfies Laman's Conditions and therefore is a 2-tree. Thus (V, E) is a 2-circuit.

Now let (V, E) be a 2-circuit and suppose that it violates one of the above conditions. Since for any edge $e \in E$, $(V, E-e)$ is a 2-tree, $|E-e| = 2|V|-3$ and $|E| = 2|V| - 2$. Hence there must be some proper subset $U \subset V$ such that $|E(U)| > 2|U| - 3$.

Let $a \in V - U$ and let $e \in E$ have a as one of its endpoints. Then, since $e \notin E(U)$, $|E(U)| > 2|U| - 3$ holds for the graph $(V, E - e)$, demonstrating that $(V, E - e)$ is not a 2-tree and contradicting the assumption that (V, E) is a 2-circuit. □

Theorem 3.26 has as a corollary a second interesting characterization of 2-circuits, the proof of which is left as an exercise.

Corollary 3.27. *A graph (V, E) is a 2-circuit if and only if there exist two trees (V, T_1) and (V, T_2) so that the following conditions hold:*

1. $T_1 \cap T_2 = \emptyset$.
2. $T_1 \cup T_2 = E$.
3. *For every* $U \subset V$, *with* $U \neq V$ *and* $|U| > 3$, *at most one of* $\big(U, T_1(U)\big)$ *and* $\big(U, T_2(U)\big)$ *is a tree.*

Exercise 3.6. *Prove this corollary.*

Exercise 3.7. *Show that both K_4 and $K_{3,4}$ are 2-circuits using the theorem and again using the corollary.*

A graph (V, E) is 2-birigid if $(V, E - e)$ is 2-rigid, for any edge $e \in E$. We have the following analogue to Theorem 2.21.

Theorem 3.28. *A graph (V, E) is 2-birigid if and only if*

1. (V, E) *is 2-rigid, and*
2. *each edge lies on a 2-circuit.*

Exercise 3.8. *Prove this theorem.*

Exercise 3.9. *Every 2-circuit is birigid. Show that the reverse is not true by constructing the simplest birigid graph that is not a 2-circuit.*

3.6 Generic Rigidity in 3-Space

We finally return to our primary interest, 3-space. Our first question is "Can we find a combinatorial characterization for 3-trees?" There are two natural places to look: an analogue to Laman's Theorem (Theorem 3.21) or an analogue to Henneberg's Theorem (Theorem 3.24). We consider the Laman approach first.

Laman's result in the plane can be described as follows. To build a 2-tree on the vertex set V, choose the correct number of edges ($|E| = 2|V| - 3$) and be sure not to waste any by packing too many among any subset of vertices,

$$|E(U)| \leq 2|U| - 3, \quad \text{for all } U \subseteq V, |U| \geq 2.$$

The 3-dimensional analogue would then read: To build a 3-tree on the vertex set V, choose the correct number of edges ($|E| = 3|V| - 6$) and be sure not to waste any by packing too many among any subset of vertices,

$$|E(U)| \leq 3|U| - 6, \quad \text{for all } U \subseteq V, |U| \geq 3.$$

We have already proved that a 3-tree must satisfy these properties (Lemma 3.20). However, these conditions are just not sufficient!

The simplest counterexample is the "double banana" graph pictured in Figure 3.22. Each "banana" consists of two tetrahedra glued together along a triangle, and the two bananas are then attached at their tips. Clearly, this framework is not rigid; the bananas can rotate independently about the axis through the tips. So it is not a 3-tree. On the other hand this graph satisfies the Laman Conditions, as we will show in a moment.

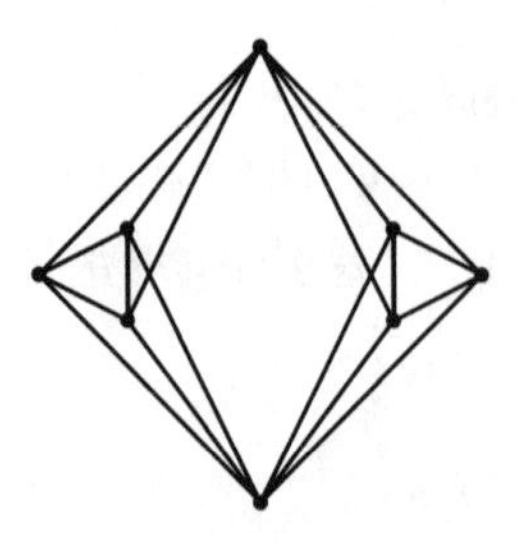

FIGURE 3.22

The following observations simplify the checking of Laman's Conditions:

Exercise 3.1. *Let (V, E) be any graph and show the following.*

1. *For all $U \subseteq V$ with $|U| = 3, 4$, $|E(U)| \leq 3|U| - 6$.*
2. *For all $U \subseteq V$ with $|U| = 5, |E(U)| \leq 3|U| - 6$, unless $(U, E(U)) = K_5$.*

In considering the double banana, (V, E), we first note that $|E| = 18$ and $|V| = 8$; so $|E| = 3|V| - 6$. Second, we note that (V, E) contains no K_5. In view of Exercise 3.1, we need only check that the inequality $|E(U)| \leq 3|U| - 6$ holds for $|U| = 6$ or 7, that is, $|E(U)| \leq 12$ when $|U| = 6$ and $|E(U)| \leq 15$ when $|U| = 7$. The number of edges in $E(U)$ when $|U| = 7$ is simply 18 minus the valence of the vertex not in U. Thus, $|E(U)| \leq 14 = 18 - 4$ when $|U| = 7$. Similarly, when deleting two vertices one must delete at least seven edges, giving $|E(U)| \leq 11$ when $|U| = 6$.

To summarize: In constructing a 3-tree, we must make sure that the Laman Conditions are satisfied, but that alone will not insure that the graph in hand is a 3-tree.

If the Laman approach will not work, perhaps the Henneberg approach will. Can we describe a method for constructing all 3-trees that is analogous to the construction of all trees and all 2-trees? Again, we have already proved part of a generalization of Henneberg's result to 3-space. Combining Lemmas 3.22 and 3.23, we have:

Theorem 3.29. *Any graph that is constructed from a triangle by a sequence of 3-extensions and 3-edge-splits is a 3-tree. Furthermore, any 3-tree with a vertex of valence 3 is a 3-extension of a smaller 3-tree, and any 3-tree with a vertex of valence 4 is a 3-edge-split of a smaller 3-tree.*

This result enables us to construct many 3-trees. As an example, we show that $K_{4,6}$ is a 3-tree. We first note that K_4 is a 3-extension of the triangle and hence a 3-tree. Now split each of the six edges of K_4, always attaching the new vertex to the vertices of the original K_4; the result is $K_{4,6}$.

Exercise 3.2. *For each of the following graphs, find a Henneberg sequence that demonstrates that it is a 3-tree.*

1. *$K_{5,5}$ minus an edge.*
2. *K_6 minus the three edges of a triangle.*
3. *K_6 minus three parallel edges (edges with no common endpoints).*
4. *$K_{4,4}$ plus any two edges (3-cases).*

However, in part 2 of Exercise 3.3, you proved that every vertex of a graph on four or more vertices that satisfies the 3-dimensional Laman Conditions has valence at least 3 and that such a graph contains at least one vertex of valence less than 6. But the possibility of a 3-tree with no vertices of valence 3 or 4 exists. Such a 3-tree could not be constructed by a sequence of 3-extensions and 3-edge-splits. Suppose that (V, E) is a 3-tree that has no vertices of valence 3 or valence 4, and let V_i denote the set of vertices of valence i for $i = 5, 6, \ldots, k$. Then

$$|V| = |V_5| + |V_6| + \cdots + |V_k|$$

and

$$|E| = \frac{5|V_5| + 6|V_6| + \cdots + k|V_k|}{2}.$$

Substituting these values into $|E| = 3|V| - 6$ and reorganizing the terms gives us the total

$$|V_5| = 12 + |V_7| + 2|V_8| + \cdots + (k-6)|V_k|.$$

Thus, a smallest graph that could be a 3-tree that has no vertices of valence less than 5 would have exactly 12 vertices, each of valence 5.

The graph of the icosahedron pictured on the left in Figure 3.23 jumps to mind. We will see in the first section of the next chapter that, early in the development of rigidity theory, the icosahedron was known to be a 3-tree. Another graph with exactly 12 vertices, each of valence 5, is $K_{6,6}$ minus six "parallel" edges (pictured to the right in Figure 3.23). This is also a 3-tree. But since all vertices are 5-valent, neither of these graphs can be constructed by a sequence of 3-extensions and 3-edge-splits.

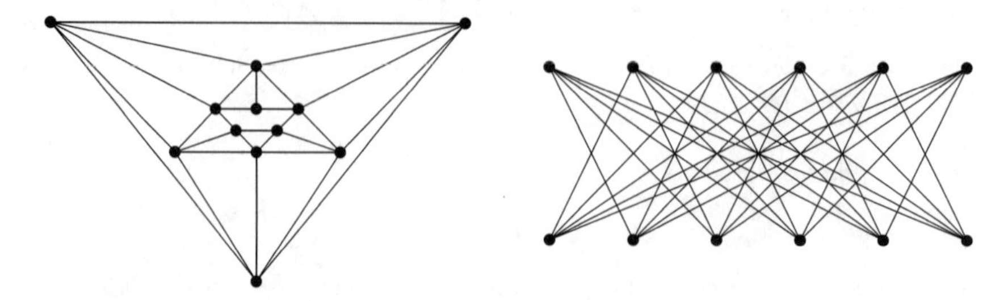

FIGURE 3.23

As we will show, these graphs pass all of the combinatorial tests for a 3-tree; but, at this point in the development of rigidity theory, we have no *purely combinatorial* way of verifying that these graphs are 3-trees.

Exercise 3.3. *Verify that both of the graphs in Figure 3.23 satisfy the Laman Conditions for a* 3*-tree.*

Suppose that (V, E) is a 3-tree and $a_0 \in V$ a vertex of valence 5, adjacent to vertices a_1, a_2, a_3, a_4 and a_5. Let $\mathbf{p}$ be a generic embedding into 3-space, and consider the framework $(V, E, \mathbf{p})$. If we delete a_0 and the five edges containing it, the resulting graph has two edges less than the number needed to be rigid. So the resulting framework has 8 degrees of freedom: the 6 degrees of freedom of a rigid body in 3-space plus 2 internal degrees of freedom corresponding to deformations of this smaller framework.

We note, as we did in the proof of Lemma 3.23, that all such deformations must alter the distance between at least one pair of points among $\mathbf{p}_1$, $\mathbf{p}_2$, $\mathbf{p}_3$, $\mathbf{p}_4$ and $\mathbf{p}_5$, otherwise it would be a deformation of our original framework. Including the edge corresponding to an altered distance results in a graph and framework with 7 degrees of freedom. Hence, by the same argument, a second edge can be added with endpoints among $\mathbf{p}_1$, $\mathbf{p}_2$, $\mathbf{p}_3$, $\mathbf{p}_4$ and $\mathbf{p}_5$ resulting in a rigid framework and a 3-tree.

The reversal of this process would be to

- delete two edges and
- attach a new vertex to five vertices including all endpoints of the deleted edges.

This is called a 3-dimensional double edge split or simply a 3*-double edge split.* We have outlined a proof of:

Lemma 3.30. *A* 3*-tree with a vertex of valence* 5 *is a* 3*-double-edge-split of a smaller* 3*-tree.*

Exercise 3.4. *Fill in the details of the proof of Lemma 3.30.* [Hint: Reread the proof of the second half of Lemma 3.23.]

It follows from Lemma 3.30 that any 3-tree can be constructed from a triangle by a sequence of 3-extensions, 3-edge-splits and 3-double-edge-splits. Unfortunately, it is not true that a 3-double-edge-split of a 3-tree is always another 3-tree!

This is illustrated in Figure 3.24. Deleting a_3 from the left-hand graph results in a $K_{4,4}$ with two additional edges. It follows from Exercise 3.2

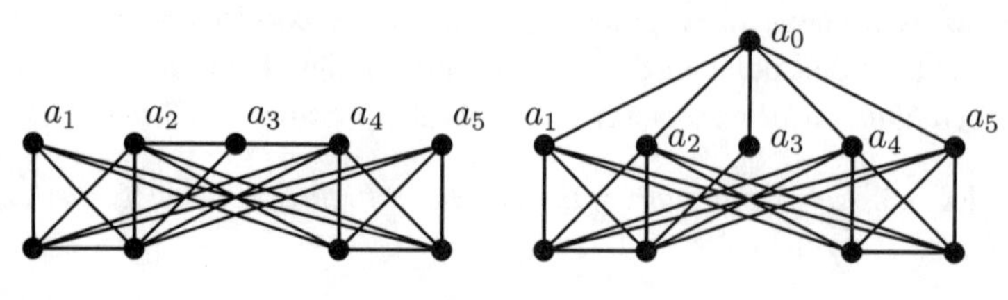

FIGURE 3.24

that this graph is a 3-tree. Thus, the 3-extension obtained by including a_3 is also a 3-tree. But, the 3-double-edge-split of the left-hand 3-tree obtained by deleting the edges $\{a_2, a_3\}$ and $\{a_3, a_4\}$ and including the edges joining a_0 to a_1 through a_5 is not a 3-tree. It has a vertex of valence 2, which is impossible in a 3-tree!

In short, we can prove that every 3-tree can be constructed by starting with a triangle and attaching a sequence of vertices of valences 3, 4 or 5 (by the methods described above). But, we cannot, at this time, prove that every graph constructed in this way is a 3-isostatic graph.

Exercise 3.5. *Show that each of the two graphs in Figure 3.23 can be constructed from a triangle by a sequence of 3-extensions and 3-edge-splits followed by a single 3-double-edge-split.*

Exercise 3.6. *Show that the double banana, Figure 3.22, cannot be constructed from a triangle by a sequence 3-extensions, 3-edge-splits and 3-double-edge-splits.*

Searching for a usable characterization of 3-trees is an area of active research. Several approaches are under investigation. One approach is to identify a class of graphs $\mathcal{C}$, including the double banana, so that one can prove the following:

> *A graph on the vertex set V with $|E| = 3|V| - 6$ edges satisfying $|E(U)| \leq 3|U| - 6$, for all $U \subseteq V$ with $|U| \geq 3$, and containing no subgraph from $\mathcal{C}$ is a 3-tree.*

Some effort has gone into this approach, and it seems now that the class $\mathcal{C}$ will be just too large and too complicated for such a theorem to be of much use, even if $\mathcal{C}$ can be characterized.

A second approach is to try to identify just when a 3-double-edge-split of a 3-tree will be a 3-tree. It is easy to see why the 3-double-edge-split in

Figure 3.24 does not work. But no one has yet been able to completely sort out just which 3-double-edge-splits work and which do not.

In spite of the fact that a complete combinatorial characterization of 3-trees has yet to be found, the steps toward such a characterization developed in this chapter will prove powerful enough for the applications we consider next, in Chapter 4.

CHAPTER 4

History and Applications

4.1 A Short History of Rigidity

In 1766, Euler conjectured:

> "A closed spatial figure allows no changes, as long as it is not ripped apart."

The spatial figures that Euler was thinking about can best be described as consisting of rigid flat polygons taped together along their edges. In more modern terminology, a closed spatial figure is a closed polyhedral surface made up of rigid polygonal plates that are hinged along the edges where plates meet.

The conjecture that all such closed polyhedral surfaces are rigid has been supported by thousands of model builders over hundreds of years. The seemingly flexible, closed polyhedral surfaces that jump to mind (chinese lanterns, accordions, bellows, etc.) all owe their flexibility to the fact that the material out of which they are made actually bends and stretches a bit. They do not contradict Euler's conjecture.

Euler was unable to prove his conjecture, and it was 47 years before the first major result toward settling his conjecture was published. As one might guess, this first result dealt with particularly nice polyhedral surfaces: the surfaces of *strictly convex polyhedra*.

- A region of the plane or of 3-space is *convex* if, for every pair of points of the region, the line segment joining them lies entirely in the region.
- A convex polyhedron is *strictly convex* if, for each vertex, there exists a plane that intersects the polyhedron in that vertex and no other point of the polyhedron.

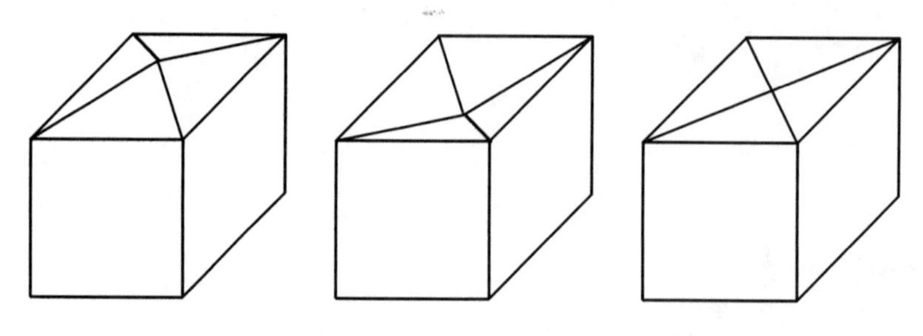

FIGURE 4.1

The first polyhedron in Figure 4.1 is strictly convex; it is constructed by putting a pyramid with its base removed on top of a cube with its top removed. The second polyhedron is constructed in the same way, except that the pyramid is pointing into the cube; it is not convex. The third polyhedron is constructed by dividing the top face of the cube into four triangles; it is convex but not strictly convex, since the top central vertex and the four triangular faces all lie in a plane.

Theorem 4.1. [A. L. Cauchy, 1813] *If there is an isometry between the surfaces of two strictly convex polyhedra that is an isometry on each of the faces, then the two polyhedra are congruent.*

Isometry is just another term for a congruence; and the gist of Cauchy's result is that, if the faces of two strictly convex polyhedra can be matched up so that incidences are preserved and corresponding faces are congruent, then the two polyhedra will be congruent. Although Cauchy's Theorem is stated in terms of congruences, it implies that a strictly convex polygonal surface is rigid: Just a slight deformation would produce a matching surface, with corresponding faces congruent, bounding another strictly convex but noncongruent polyhedron.

It is easy to see that the convexity condition cannot be dropped from the hypothesis of Cauchy's Theorem. Consider the first two polyhedra in Figure 4.1. There is an obvious matching between these two surfaces, so that corresponding faces are congruent; but the solid polyhedra are clearly not congruent. But why strict convexity? Consider the third polyhedron in Figure 4.1. This polyhedron is convex but not strictly convex. Nevertheless, our intuition tells us that this polyhedron is rigid, as indeed it is. However, it is not infinitesimally rigid!

What do we mean by an infinitesimal motion of a polyhedral surface? We mean an assignment of vectors to the vertices so that the projections of the vectors at the vertices of a face into the plane of that face form an infinitesimal rigid motion of that face. If we assign the zero vector to all of the vertices of the cube and a vertical unit vector to the apex of the flat pyramid, we have an infinitesimal motion of this polyhedron. If we were to develop infinitesimal rigidity for polyhedral surfaces a bit further, we would easily see that this infinitesimal motion of the polyhedron is not an infinitesimal rigid motion of the polyhedron. Although the concept of infinitesimal rigidity had not yet been formulated, it seems as if Cauchy found it easier to work with polyhedra that were not just rigid but infinitesimally rigid as well.

These spatial figures of Euler and Cauchy are now called *plate and hinge frameworks:* frameworks with flat rigid polygonal plates for faces, joined by flexible hinges along their edges. To each such plate and hinge framework we can associate a rod and joint framework: We replace the hinges by rods, replace the vertices by joints and delete the plates. This associated rod and joint framework is called the 1-*dimensional skeleton* (or simply the 1-*skeleton*) of the plate and hinge framework. Since every triangle is rigid in 3-space, a triangular plate and its 1-skeleton are equally rigid. So for a plate and hinge framework $\mathcal{F}$ with all triangular faces and its 1-skeleton $\mathcal{S}$, either $\mathcal{F}$ and $\mathcal{S}$ both will be rigid or neither of them will be rigid.

In Figure 4.2, we have pictured an octahedron and the structure graph of its 1-skeleton. Any rod and joint framework in 3-space with this structure graph is called an octahedral rod and joint framework; any plate and hinge framework with an octahedral rod and joint framework as its 1-skeleton is called an octahedral plate and hinge framework.

A complete study of octahedral rod and joint frameworks and octahedral plate and hinge frameworks was carried out by R. Bricard and constitutes the next major step in the investigation of plate and hinge frameworks. The main result of that study can be stated as follows:

Theorem 4.2. [R. Bricard, 1897] *Every octahedral plate and hinge framework is rigid. However, there are octahedral rod and joint frameworks in 3-space that are not rigid.*

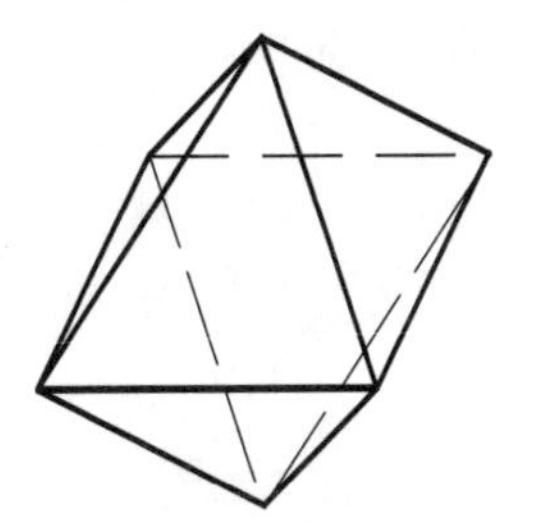
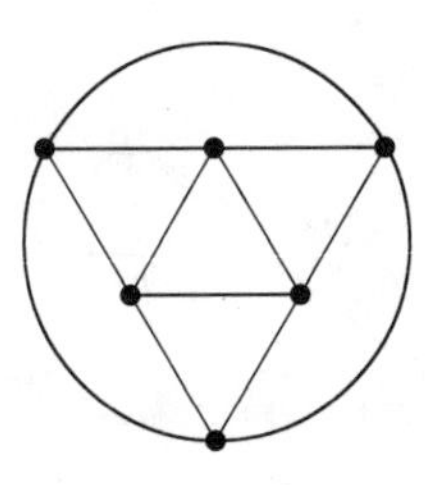

FIGURE 4.2

At first glance the two parts of Bricard's theorem seem to contradict one another. Since the faces of the octahedron are all triangular, deleting them could not change a rigid plate and hinge framework into a nonrigid rod and joint framework. However, this seeming contradiction vanishes once we note that not all octahedral rod and joint frameworks in 3-space are the 1-skeletons of plate and hinge frameworks. A natural first step in analyzing Bricard's result is to check whether the octahedral map satisfies the Laman Conditions (Lemma 3.20). This is straightforward and is left as an exercise for the reader.

Exercise 4.1. *Verify that the octahedral map satisfies the 3-dimensional Laman Conditions.*

The nonrigid octahedral rod and joint frameworks discovered by Bricard are not too difficult to describe. Consider a nonregular pyramid with a flat square base, and assume that the perpendicular projection of the apex onto the base does not lie on any line of symmetry of the square. See the left-hand framework in Figure 4.3. Note that this pyramid is not rigid; the base easily distorts once it is no longer required to be flat.

The first step in understanding the deformations of this figure is compute its internal degrees of freedom: five vertices, eight edges gives $(3\times5-6)-8=1$ internal degree of freedom for this framework. To understand the deformations of this framework, visualize stretching (or shrinking) the distance between the points labeled $\mathbf{p}$ and $\mathbf{q}$. This will force the apex of the pyramid to move down (or up); just how the remaining two vertices of the base move depends on the relative lengths of the four rods at the apex. We should point out that only the base is deformed by this motion; the four triangles are each rigid and remain undeformed. We need to understand just how the base can be deformed.

Consider the base alone as a framework. It has 2 internal degrees of freedom: $(3 \times 4 - 6) - 4 = 2$. Visualize a deformation of this framework in 3-space. Next visualize inserting a temporary edge between the $\mathbf{p}$ and $\mathbf{q}$. With the additional edge the framework has 1 degree of freedom. This is easy to see, since it now consists of two isosceles triangles with a common base;

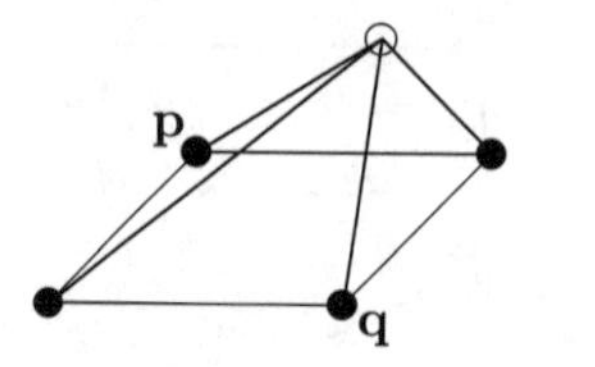

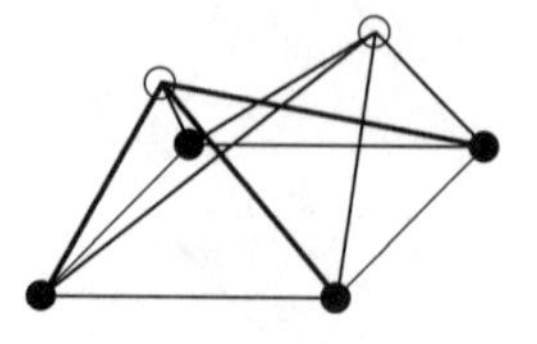

FIGURE 4.3

the remaining degree of freedom is represented by the freedom to choose the angle that the planes of the isosceles triangles make with one another. When the apex is reattached, it will determine this angle.

This way of looking at the deformed square base makes it clear that the base, under any deformation, has a symmetry: the 180-degree rotation about the line that is a perpendicular bisector of the temporary edge and that also bisects the angle between the isosceles triangles. We now use this observation to build Bricard's flexible octahedra.

Start with the undeformed pyramid and construct a second pyramid on the *same* base that is congruent to the first under the 180-degree rotation about the line perpendicular to the base at its center. Note that both apexes are on the same side of their common base. See the right-hand framework in Figure 4.3. It follows from our observation about the symmetry of the deformed base that both pyramids produce the same deformation of the base as they flex. Hence, this octahedral rod and joint framework is nonrigid. Finally, note that filling in the triangular faces of this joint and rod framework is impossible. In other words, this joint and rod framework is not the 1-skeleton of a plate and hinge framework.

The Euler Conjecture was finally settled in 1977. Just prior to that, a result was published that greatly extended the set of configurations for which the Euler Conjecture was known to be valid:

Theorem 4.3. [H. Gluck, 1975] *Every closed, simply connected, polyhedral surface generically embedded in 3-space is rigid.*

Gluck's Theorem tells us that the Euler Conjecture is almost always true for closed, simply connected, polyhedral surfaces. A *simply connected surface* is a sphere-like surface. A doughnut-shaped polyhedral surface, for example, is not simply connected.

Just two years after Gluck proved his theorem, a counterexample to the Euler Conjecture was constructed by Robert Connelly. The counterexample is based on Bricard's flexible octahedron and, of course, is not strictly convex; in fact, it is not convex. Since it flexes, it must be nongeneric (by Gluck's Theorem). So, a model of this polyhedral surface must be accurately constructed in order to flex. Precise instructions for constructing a model of a simplification of Connelly's polyhedron (due to Klaus Steffen) are included in "Mathematical Recreations" by A. K. Dewdney in the May 1991 issue of *Scientific American* magazine.

We have followed the progress of the investigation of Euler's Conjecture over more than 200 years. In the next section, we follow the thread of another interesting and important investigation that arose out of practical applications.

4.2 Linkages and Curves

Consider planar rod and joint frameworks that have some joints pinned. If a joint is pinned it cannot move, but the rods at that joint can rotate about that joint. In the literature, these are called *linkages.* The simplest linkage consists of a single rod pinned at one end. This linkage has 1 degree of freedom:

- 0 degrees of freedom, for the pinned joint;
- plus 2 degrees of freedom, for the free joint;
- minus 1 degree of freedom, for the segment.

If we put a pen at the free endpoint, the pen traces a circle under the motion of this linkage. The circle is a 1-dimensional curve, corresponding to the 1 degree of freedom of the linkage.

We actually use a version of this linkage, a pair of compasses, to draw our circles. By contrast, we draw straight lines using a template—a straightedge. The analogous way to draw a circle would be to trace around the edge of an appropriately sized disk. These observations lead to a natural question:

Is there a simple linkage that traces a straight line segment?

This question, like Euler's conjecture, remained unsettled for an extended period of time.

The history of this problem and the early history of the theory of linkages is beautifully described by A. B. Kempe in "How to Draw a Straight Line." Originally a lecture given to a group of science teachers in the summer of 1876, this fascinating little book was the main resource used in the writing of this section.

Before we tackle the straight line linkage problem, let's develop a bit of the general theory of linkages. The next simplest linkage, after a single rod, consists of two rods joined at a common endpoint with one of their joints pinned. Suppose that the pinned joint is not the common joint. Let $\mathbf{p}_0$ denote the pinned joint; let $\mathbf{p}_1$ denote the joint linking the two rods and let $\mathbf{p}_2$ denote the free endpoint. This linkage has 2 degrees of freedom,

$$2 = 0 + 2 + 2 - 1 - 1,$$

from $\mathbf{p}_0$, $\mathbf{p}_1$, $\mathbf{p}_2$ and the two segments, respectively; see Figure 4.4. If we attach a pen at $\mathbf{p}_2$, it will fill a portion of the plane under the motions of the linkage. If the rod joining $\mathbf{p}_1$ to $\mathbf{p}_0$ is the longer rod, we get an annulus; if the rod joining $\mathbf{p}_1$ to $\mathbf{p}_0$ is the shorter rod, a disk is filled in.

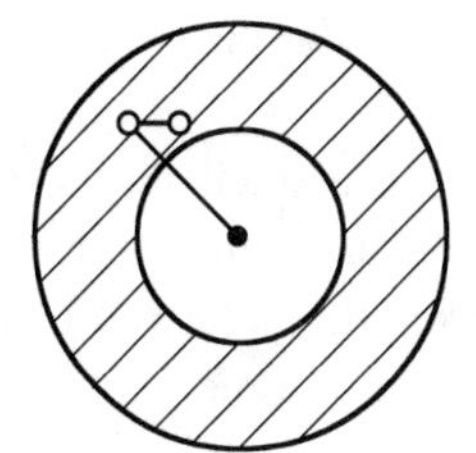
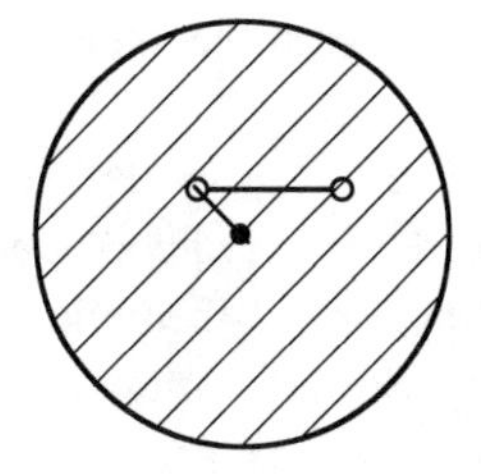

FIGURE 4.4

Exercise 4.1. *Analyze the simple two-segment linkage with the common joint pinned.*

We are interested in curves, so we want linkages with 1 degree of freedom. The degrees of freedom accounting goes as follows:

- each pinned joint contributes 0 to the degrees of freedom count;
- each unpinned joint contributes 2; and
- each nonredundant rod decreases the degrees of freedom count by 1.

So a linkage with u unpinned joints and r, properly placed, rods has $2u - r$ degrees of freedom. Setting $2u - r$ equal to 1 and solving gives: $r = 2u - 1$. Thus, the linkages of interest to us have an odd number of rods. The simplest $(r = u = 1)$ is the circle drawing linkage. The next simplest curve drawing linkages will have three rods.

There are several 3-rod linkages of historical interest. Two of them are pictured in Figure 4.5. The larger black dots denote pinned joints; the white dots represent (unpinned) joints; and the smaller black dot represents the pen point. Of course, these linkages have the same structure graph. However, the embedding has a dramatic effect on the curves they draw. The first linkage in the figure was invented in 1784 by James Watt and is called "Watt's parallel motion"; the second linkage was designed about 50 years later by a well-known Russian mathematician, Pafnuti Tchebycheff. There is no vertex

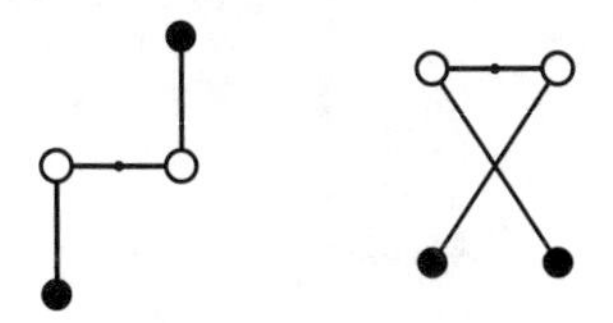

FIGURE 4.5

where the rods cross in the Tchebycheff linkage and the two rods slide across one another.

The interest in these two linkages is that, near their initial positions, the curves they trace are very good approximations of a straight line. The interest was practical as well as theoretical, for during the industrial revolution such linkages were used in a variety of mechanisms. In a mechanical device, such a linkage could replace a peg sliding in a straight slot, resulting in a smoother motion and a reduction in friction.

One can easily build models of the linkages pictured in Figure 4.5 and actually trace the curves to see just how well they approximate a straight line segment. For either of the linkages, a flat smooth 12-inch by 12-inch board, a sheet of paper, several thumbtacks, three strips of stiff cardboard, a sharp pencil and an ice pick or other such sharp tool are all that is needed to construct a working model. The unpinned joints are made with thumbtacks pointing up and the ice pick is used to punch the hole for the pencil point.

Models for the Watt and Tchebycheff linkages are pictured in Figure 4.6. Very nice geometric software now available, *Geometer's Sketchpad* and *Cabri*, enables one to model these linkages on the computer. Both are available in Windows and Macintosh versions; *Cabri* is also on the TI-92 calculator.

In 1868, a French engineer named Peaucellier produced a linkage that drew a straight line segment. The fundamental part of the mechanism, called

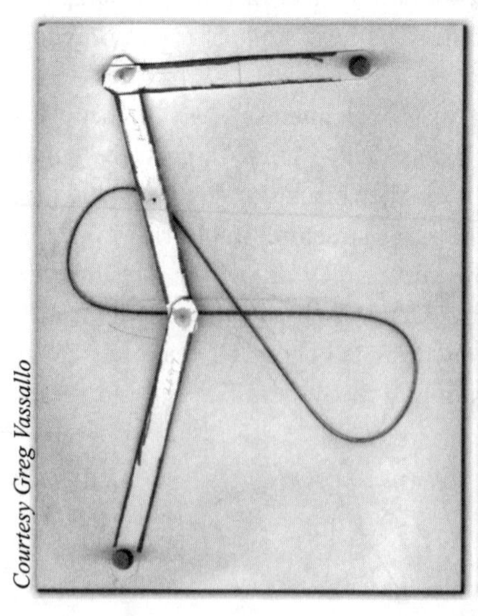

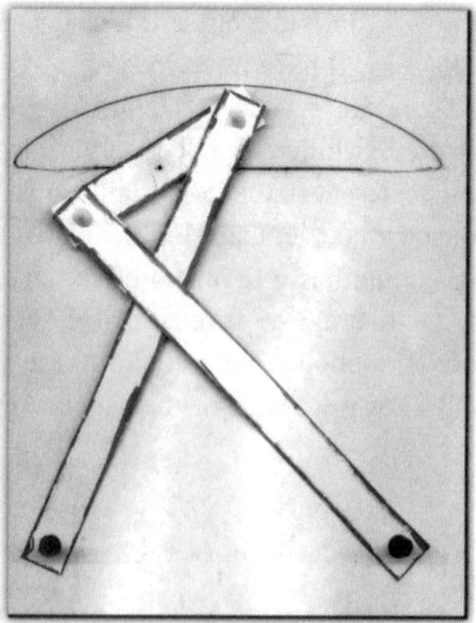

Courtesy Greg Vassallo

FIGURE 4.6

Peaucellier's cell, is pictured to the left in Figure 4.7. One easily checks that the full Peaucellier linkage, pictured to the right in the figure, has 1 degree of freedom ($u = 4$ and $r = 7$). So, the pen at joint Q will trace a curve of some sort.

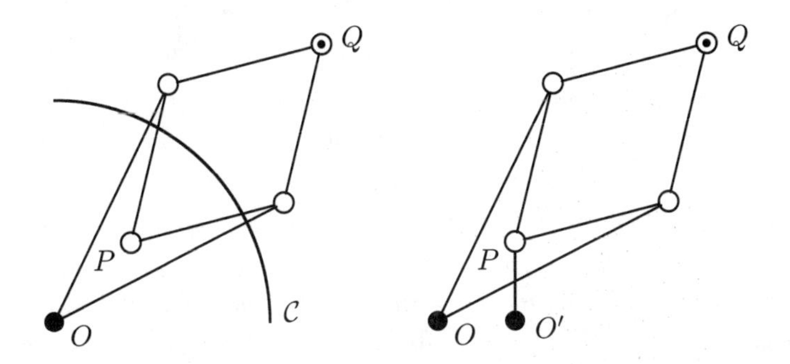

FIGURE 4.7

The proof that the trajectory of Q is indeed a line segment has two parts. The main part has to do with Peaucellier's cell. This linkage has 2 degrees of freedom. Suppose that the cell is deformed so that P and Q coincide; let $\mathcal{C}$ denote the circle traced by the pen when the cell, in this configuration, is rotated about O. Once the circle has been traced, let P and Q move relative to one another again. The critical feature of Peaucellier's cell can now be stated:

> *The points P and Q are inverses of one another with respect to the circle $\mathcal{C}$.*

(Points P and Q are inverses of one another with respect to the circle with center O and radius c if O, P and Q are collinear and $|OP||OQ| = c^2$.) For Peaucellier, the critical feature of circle inversion is this:

> *Circles through the origin invert to straight lines!*

Thus to complete his device he simply added a pinned joint O' and a link between it and P of length equal to the distance between O' and O. This forces P to move along a circle through O, and its inverse image Q must then move along a straight line!

In the next exercise, we lead the interested reader through a detailed analysis of Peaucellier's linkage.

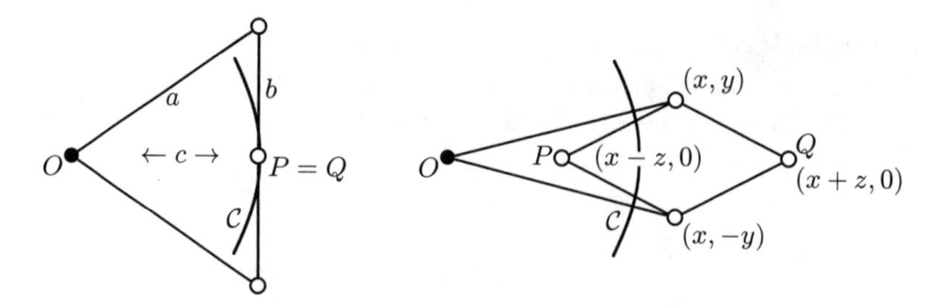

FIGURE 4.8

Exercise 4.2. *Consider Peaucellier's Cell as redrawn on the right in Figure 4.8 with the pinned vertex O at the origin. Note that by symmetry O, P, Q are collinear. Choose, for the positive x-axis, the ray from O that passes through P and Q. Denote by $\mathcal{C}$ the circle with center at O and radius c, equal to the distance from O to P when P and Q coincide. Let a denote the length of the cell's two longer rods while b denotes the length of the four shorter rods.*

1. *Prove that P and Q are inverses with respect to C.*

   (a) *From the left-hand drawing in Figure 4.8, establish the equation $c^2 = a^2 - b^2$.*

   (b) *Verify that, in the given coordinate system, when the upper joint is labeled (x, y) the remaining joints must be labeled as depicted on the right in Figure 4.8. In particular, verify the equations $|OP| = x - z$ and $|OQ| = x + z$, for some number z.*

   (c) *Next, verify the equations $x^2 + y^2 = a^2$ and $z^2 + y^2 = b^2$.*

   (d) *Combining these five equations, show that $|OP||OQ| = c^2$.*

2. *Prove that circles through the origin invert to straight lines.*

   (a) *Keeping O at the origin, let Peaucellier's Cell move freely; in other words, do not constrain P and Q to lie on the x-axis. Letting $(\hat{x}, \hat{y})$ denote the coordinates of Q, show that the coordinates of P are $\left(c^2\hat{x}/(\hat{x}^2 + \hat{y}^2), c^2\hat{y}/(\hat{x}^2 + \hat{y}^2)\right)$. Note first that since O, P and Q are collinear, P has coordinates $(m\hat{x}, m\hat{y})$. So you can use the equation $|OP||OQ| = c^2$ to show that $m = c^2/(\hat{x}^2 + \hat{y}^2)$.*

   (b) *Verify that the circle through the origin with center at (s, t) is given by the equation $x^2 - 2sx + y^2 - 2ty = 0$.*

(c) *Substitute $c^2\hat{x}/(\hat{x}^2+\hat{y}^2)$ for x and $c^2\hat{y}/(\hat{x}^2+\hat{y}^2)$ for y in the equation of the circle. Simplify, thereby showing that, if P is constrained to move on this circle, Q is constrained to move along the straight line with equation $c^2 - 2sx - 2ty = 0$.*

Prior to the presentation of his now-famous lecture, Kempe had shown that any algebraic curve in the plane can be traced (perhaps one piece at a time) by a suitable linkage. Another more recent use of linkages has been explained in *Scientific American* by William Thurston and Jeffrey Weeks. In their article, "The Mathematics of Three-Dimensional Manifolds," Thurston and Weeks use linkages to help one visualize and study 2- and 3-dimensional surfaces. We give just one simple example here.

Consider the two-segment linkage pictured to the left in Figure 4.9. Again the pinned joint O is indicated by a black dot and the unpinned joints P and Q by white dots. It has 2 degrees of freedom. Thurston and Weeks identify each position that this linkage may take as a point on an abstract 2-dimensional surface or manifold. Intuitively, it is clear what we mean when we say that two positions are "near to one another," and this concept of "nearness" gives the topological structure of the surface.

To get a better idea of what this surface is, we coordinatize it. Note that the position of the linkage is entirely determined by the counterclockwise angle each segment makes with the horizontal. Thus, using radian measure, each position may be identified with a unique point in the $2\pi \times 2\pi$ square on the right in Figure 4.9. Since the angles with measure 0 and 2π are the same, the right and left sides of this square must be identified, as must the top and bottom. With these identifications, we see that the surface represented is the torus.

Another way of looking at this surface is to concentrate on the endpoint P. As we have already noted, P can assume any position in an annulus centered at O. We note that for every position of P interior to the annulus there are two distinct positions of the entire linkage: Reflect the entire linkage about the line through P and O. Hence, we may think of the surface as two annuli whose inner and outer rims are identified.

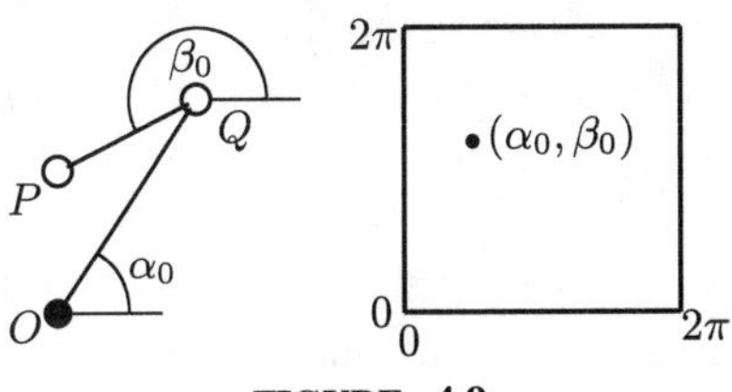

FIGURE 4.9

Exercise 4.3. *Attach a third segment (P, R) at P to the linkage in Figure 4.9.*

1. *Verify that one gets a linkage with 3 degrees of freedom.*
2. *Give a correspondence between the positions of this linkage and the points of a cube with opposite sides identified: the 3-torus.*

After working through these simple examples, Thurston and Weeks use linkages to consider more complicated 2- and 3-dimensional manifolds. The reader will be able to verify that the linkage pictured in Figure 4.10 has 2 degrees of freedom. In their paper, Thurston and Weeks show that the surface generated by this linkage is a 3-hole torus.

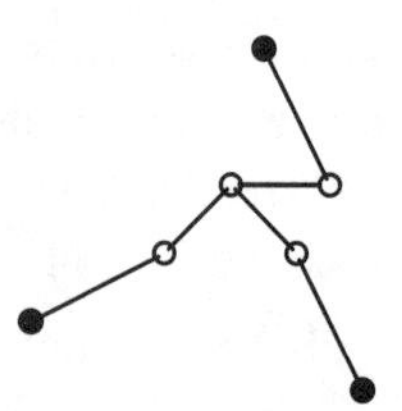

FIGURE 4.10

4.3 Triangulated Surfaces and Geodesic Domes

Some of the most important plate and hinge frameworks are those that are used to roof over structures or to completely enclose a space. We shall call such frameworks *domes.* Virtually all domes that have actually been constructed employ rather light plates attached to an already rigid rod and joint framework, its 1-skeleton. So we are mainly interested in domes with rigid 1-skeletons; in fact, we are really interested in the 1-skeletons themselves.

Such a 1-skeleton is a rigid, 3-dimensional rod and joint framework. But not just any rigid, 3-dimensional rod and joint framework—we have to be able fill in the plates. In addition to the information contained in the structure graph (V, E), we should indicate where the plates are to go. To accomplish this we attach to every dome a *structure map,* (V, E, F). The vertices and edges of this map form the structure graph for the 1-skeleton, and the faces indicate the placement of the plates. With this definition, the large body of results about planar graphs is available for the investigation of domes. The reader may wish to review the results included in Section 2.7 before going any farther.

Courtesy Greg Vassallo

FIGURE 4.11

The two types of domes that we will consider are those that are used to roof over structures and those that completely enclose a space. Examples of the former are pictured in Figure 4.11: three domes covering three water towers near the author's home. Perhaps the most famous example of the latter is the geodesic dome designed by Buckminster Fuller and built for the 1959 World's Fair in Montreal. A dome that is used as the roof of a structure has one large outside face and the joints around this face are pinned or attached to the structure. We will consider this class of domes later in this section. To start our investigation, we consider domes that, like Buckminster Fuller's geodesic domes, completely enclose a space. These are the plate and hinge frameworks first considered by Euler and Cauchy, the surfaces of polyhedra. However, we are interested only in those that have rigid 1-skeletons.

Consider a dome with structure map (V, E, F). Its 1-skeleton can be infinitesimally rigid only if it satisfies Laman's Condition, that is, only if $|E| \geq 3|V| - 6$ (see Corollary 3.10). It is natural to ask whether a map can actually satisfy this inequality. The answer is "yes—but with no room to spare," as we now show.

Lemma 4.4. *Let (V, E, F) be a map, Then $|E| \leq 3|V| - 6$ with equality if and only if all of the faces are triangular.*

Proof. Recall Euler's formula, Theorem 2.27: $|V| - |E| + |F| = 2$. By Lemma 2.26, $\hat{\rho}|F| = 2|E|$, where $\hat{\rho}$ denotes the average face valence. Solving this equation for $|F|$ and substituting it into Euler's formula gives

$$|V| - |E| + \frac{2|E|}{\hat{\rho}} = 2.$$

Solving the resulting equation for $|E|$ gives

$$|E| = \frac{\hat{\rho}}{\hat{\rho} - 2}(|V| - 2).$$

We note that $\hat{\rho} \geq 3$ with equality only if every face is a triangle. We also note that, for $\hat{\rho} \geq 3$, $\hat{\rho}/(\hat{\rho} - 2)$ is a strictly decreasing function in $\hat{\rho}$. Hence

$$|E| = \frac{\hat{\rho}}{\hat{\rho} - 2}(|V| - 2) \leq 3|V| - 6,$$

with equality if and only if $\hat{\rho} = 3$, that is, every face is a triangle. □

We have shown, in Lemma 3.19, part (3), that every 3-tree (V, E) must satisfy the equation $|E| = 3|V| - 6$. The converse is, of course, false; there are many graphs (V, E) satisfying this equation that are not 3-trees. However, for planar graphs, the converse does hold!

Theorem 4.5. *Every planar graph (V, E) with $|E| = 3|V| - 6$ is a 3-tree.*

Proof. Let (V, E) be a planar graph with $|E| = 3|V| - 6$. Then, as we have just shown, all of its faces are triangles. Draw this graph on the surface of a sphere in 3-space in such a way that the triangular faces may actually be replaced by flat triangles. It is not too difficult to show that the resulting polyhedral surface is strictly convex. Thus, by Cauchy's Theorem 4.1, it is rigid as a plate and hinge framework. But the faces, being triangular, may be deleted, leaving its 1-skeleton rigid. □

The most prominent and extensively studied domes are the *geodesic domes.* The term geodesic dome was used by Buckminster Fuller to mean more than simply a triangulated polyhedral plate and hinge framework. The additional conditions implied by the term geodesic dome include:

- the vertices should lie on, or at least be close to, a sphere;
- the triangles should be as close to equilateral as possible.

The three platonic solids, the tetrahedron, the octahedron and the icosahedron, satisfy both of these conditions exactly. Except for these three, it is impossible to construct geodesic domes out of congruent equilateral triangles and have all vertices on a sphere. The problem arises with vertices of valence 6 or more. For example, a polyhedral surface constructed with equilateral triangular faces would be flat at a vertex of valence 6, preventing all of the vertices from lying on a sphere. Vertices of higher valence cause even more problems. Hence, geodesic domes usually avoid vertices of valence greater than 6.

On the other hand, vertices of valence 3 or 4 require greatly distorted triangular faces when incorporated in geodesic domes with large numbers of faces. Since the plates to fill in the triangular faces are to be made out of lightweight material, the faces themselves cannot be too large. Thus, large geodesic domes have large numbers of faces and employ only vertices of valences 5 and 6. We will restrict our discussion to such geodesic domes, leaving the considerations of domes with other valences to the exercises.

Theorem 4.6. *Let (V, E, F) be the structure map of a geodesic dome with all vertex valences equal to* 5 *or*6. *Then, with h denoting the number of vertices of valence* 6, *we have:*

- *(V, E, F) has exactly* 12 *vertices of valence* 5*;*
- *(V, E, F) has exactly $12 + h$ vertices;*
- *(V, E, F) has exactly $30 + 3h$ edges; and*
- *(V, E, F) has exactly $20 + 2h$ faces.*

Proof. Let x denote the number of vertices of valence 5. Then $|V| = x + h$; and, by Lemma 2.1, $|E| = (5x + 6h)/2$. Substituting these values for $|V|$ and $|E|$ into $|E| = 3|V| - 6$, h cancels out and we have $x = 12$. Substituting 12 for x in the above formulas for $|V|$ and $|E|$ gives the formulas for the number of vertices and edges. Finally we observe that, since all faces are triangular, $3|F| = 2|E|$ and the formula for the number of faces follows. □

Exercise 4.1. *Fill in the blanks and prove the following Theorem. Then verify that the octahedron is the simplest dome of this type.*

Theorem 4.7. *Let (V, E, F) be the structure map of a geodesic dome with all vertex valences equal to* 4 *or* 6. *Then, with h denoting the number of vertices of valence* 6, *we have:*

- *(V, E, F) has exactly* ____ *vertices of valence* 4*;*
- *(V, E, F) has exactly* ____ *$+ h$ vertices;*

- *(V, E, F) has exactly* ____ $+$ ____h *edges; and*
- *(V, E, F) has exactly* ____ $+$ ____h *faces.*

The geodesic dome with the smallest structure map is the icosahedron. It has no 6-valent vertices and, by the previous theorem, it has 12 vertices, all of which are 5-valent, 20 faces and 30 edges. The next smallest possibility would have just one 6-valent vertex for 13 vertices, 33 edges and 22 faces. As it turns out, no such geodesic dome exists.

Exercise 4.2. *Show the following.*

1. *There exists no geodesic dome on* 13 *vertices.* [Hint: Start with one vertex of valence 6, complete the triangular faces around it and continue to work out, using only vertices of valence 5, until the graph is complete.]
2. *There is exactly one geodesic dome on* 14 *vertices.*

The geodesic domes that have been studied most extensively are those with a high degree of symmetry. A geodesic dome with a high degree of symmetry is usually constructed by starting with the icosahedron and subdividing each of the faces in the same way. In Figure 4.12, we illustrate three possible subdivisions of an icosahedral face. The original 5-valent vertices are represented by the larger dots; the smaller dots represent the new, 6-valent vertices. In Figure 4.13, we have a picture of an icosahedral model with its faces subdivided by the scheme on the left in Figure 4.12.

It is convenient to index the subdivisions like those in Figure 4.12 by n, the number of edges into which each original edge has been subdivided. Thus I_1 denotes the icosahedron; I_2, the icosahedron with faces subdivided as on the left in the figure; and so on. Each I_n will of course have 12 5-valent vertices. Instead of computing h, the number of vertices of valence 6 directly, we compute the number of faces. We note that the number of faces of I_n in one face of the superimposed icosahedron is $1 + 3 + \cdots + (2n - 1) = n^2$.

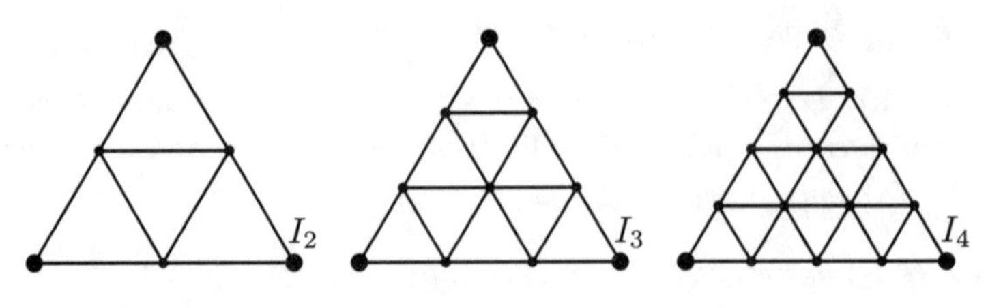

FIGURE 4.12

FIGURE 4.13

Thus I_n has $20n^2$ faces. Using Theorem 4.6, we compute

$$h = \frac{20n^2 - 20}{2} = 10(n^2 - 1),$$

compute the total number of vertices to be $10n^2 + 2$, and compute the number of edges to be $30n^2$.

Exercise 4.3. *Check the algebra in the above discussion. In particular, prove that the sum of consecutive odd numbers* $1 + 3 + \cdots + (2n - 1)$ *does equal* n^2.

As is apparent in Figure 4.13, if you construct an actual model of I_n using equilateral triangles, that model will not be spherical at all: It will be just an icosahedron with its faces subdivided but still flat. In constructing geodesic domes, we carry out this subdivision process on the sphere:

1. First construct the circumsphere of the icosahedron, and replace each edge with the arc of the great circle through its endpoints.
2. Subdivide these arcs into n equal subarcs, and complete the subdivision of the face on the corresponding region of the sphere using great circle

arcs for the edges. The result will be a copy of I_n with all vertices on the sphere and all edges represented by arcs on the sphere.

3. Finally, replace each arc representing an edge by a straight line segment.

The resulting model of I_n will have all of its vertices on a sphere, but its edges will have differing lengths and its faces will no longer be equilateral triangles. To actually build such a model of I_n, one must compute the lengths of the edges. This is not so easy to do. These computations have been made and published in the many books and articles on geodesic domes. Several are listed at the back of this book.

There is a relatively easy way to compute the lengths for a model of I_n that approximates the spherical model. The key observation used in developing this model is this:

The curvature of the dome at a vertex can be roughly represented by the sum of the angles around that vertex.

The idea is to make the curvature at every vertex approximately the same by choosing the angles of the triangular faces so that the sum of the angles around a vertex is independent of the choice of vertex. This will lead to a system of linear equations that admits many solutions. We will opt for those solutions that exhibit the highest degree of symmetry.

For example, consider I_2. Take any one of the faces with a 5-valent vertex, and let x denote the measure (in degrees) of the angle at the 5-valent vertex. Assuming that the subdivided icosahedral face is symmetric about its center, the remaining two angles of the face are equal to each other and, hence, equal to $(180 - x)/2$ degrees. Also, by symmetry, the central triangle is equilateral. See the left-hand diagram of Figure 4.14. The condition that the sum of the

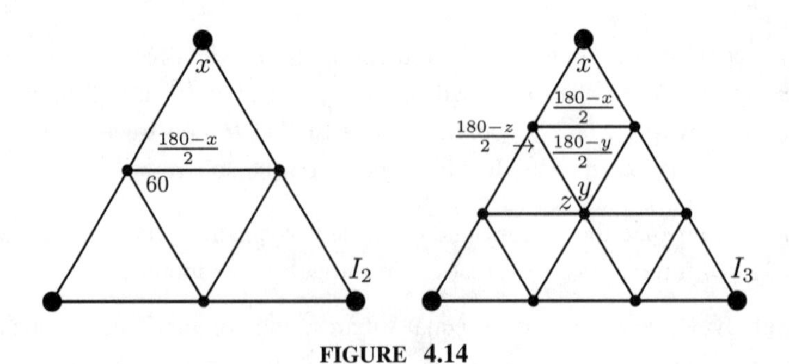

FIGURE 4.14

angles around a vertex is the same at each vertex is

$$5x = 4\left(\frac{180 - x}{2}\right) + 120.$$

Solving gives $x = \frac{480}{7}$ or approximately 68.8 degrees. The other two vertices of the top triangle are $\frac{390}{7}$ or approximately 55.7 degrees. Using the Law of Sines, we may compute the ratio of the lengths of the two types of edges. Taking the lengths of the edges joining a vertex of valence 5 and a vertex of valence 6 to be 1, the lengths of the edges joining two vertices of valence 6 is

$$\frac{\sin\left(\frac{480}{7}\right)}{\sin\left(\frac{390}{7}\right)}$$

or approximately 1.13.

Exercise 4.4. *Again assume a high degree of symmetry, and work out a set of angles and lengths for* I_3.

1. *Verify that the system of equations for this case is*

$$5x = 3y + 3z \qquad \text{and}$$

$$5x = 2\left(\frac{180 - x}{2}\right) + 2\left(\frac{180 - y}{2}\right) + 2\left(\frac{180 - z}{2}\right).$$

2. *Verify that the solution to this redundant system is*

$$x = \frac{1620}{23} \approx 70.4 \text{ degrees} \qquad \text{and}$$

$$y + z = \frac{2700}{23} \approx 117.4 \text{ degrees}$$

3. *Taking the edges joining a* 5*-valent and* 6*-valent vertex to have length* 1*, use the Law of Sines to to show that the edges opposite the* 5*-valent vertices have length approximately* 1.15 *units.*

4. *Now argue that, if* $y < 60$*, then the edges adjacent to the central vertex would be longer than* 1.15 *units. If* $y = 60$ *degrees, then the edges adjacent to the central vertex would have length* 1.15 *units,* z *would equal* 57.4 *degrees and the edges opposite the* z*-angles would be somewhat shorter than* 1.15 *units. To avoid these edges getting too small, take* $y = 60$.

5. *Compute* z.

6. *Compute the lengths of the remaining edges.*

Consider any geodesic dome with h 6-valent vertices, and assume that the sum of the measures of angles around each vertex is the constant s. Sum these sums over all vertices to get $s(12+h)$. Now let's compute this sum another way:

Sum the angles around each face and then sum over all faces.

Thus,

$$s(12+h) = 180(20+2h) = 3600 + 360h \text{ degrees.}$$

We conclude that the sum at each vertex is

$$\frac{3600+360h}{12+h} \text{ degrees.}$$

Rewriting the numerator as $4320 + 360h - 720$, we compute the sum at each vertex to be

$$360 - \frac{720}{12+h}.$$

If we reconsider I_2, where $h = 30$, we have at once that

$$5x = 360 - \frac{720}{42} = \frac{2400}{7} \quad \text{and} \quad x = \frac{480}{7}.$$

Exercise 4.5.

1. *Use* $360 - [720/(12+h)]$ *for the sum at each vertex to simplify the computations of the angles for the cases of* I_3.

2. *Assume a high degree of symmetry and work out a set of angles and lengths for* I_4. [Hint: To start with, use arguments given in our investigations of I_2 and I_3 to decide which faces should be equilateral triangles.]

Exercise 4.6. *Let the map* (V, E, F) *be given, and consider any polyhedron with it as structure map. Prove that the sum of the sums of the angles around each vertex is simply* $360|V| - 720$ *degrees.* [Hint: Show first that the sum of the angles of a polygon with k edges is $180k - 360$ degrees. Then sum over all faces and plug into Euler's Formula.]

In our notation, I_1 is the icosahedron and has no vertices of valence 6; I_2 introduces a vertex of valence 6 in the center of each edge of the icosahedron, resulting in 30 6-valent vertices. There is a different kind of subdivision with only 20 6-valent vertices:

1. Start with the icosahedron and add a new vertex in the center of each face.
2. Join each new vertex to the three vertices of the face containing it.

We now have an embedded graph with 60 triangular faces, 12 10-valent vertices and 20 3-valent vertices.

3. Remove the edges joining two 10-valent vertices.

The result is an embedded graph with 30 quadrilateral faces, 12 5-valent vertices and 20 3-valent vertices.

4. Finally, put an edge across each of these quadrilateral faces joining the two 3-valent vertices in its boundary.

The result is a symmetric embedded graph with 60 triangular faces, 12 5-valent vertices and 20 6-valent vertices. A model of this geodesic dome is pictured in Figure 4.15.

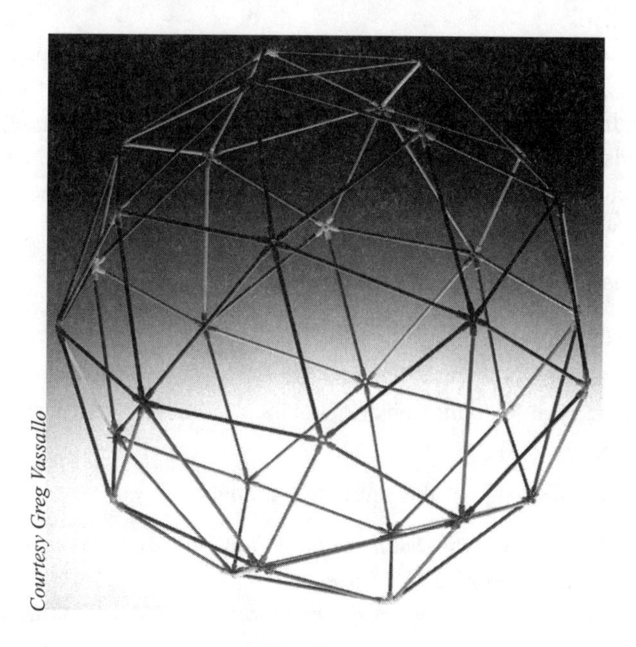

FIGURE 4.15

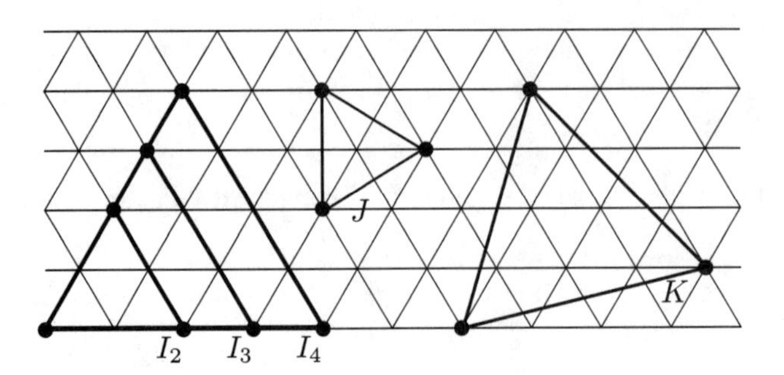

FIGURE 4.16

Exercise 4.7. *Compute the angles and edge ratios of the geodesic dome just described, assuming symmetry about each vertex and equal angle sums at each vertex.*

For those studying geodesic domes, finding, cataloging and comparing all methods of subdivision is the "name of the game." The easiest way to visualize the different methods of subdivision is to consider the tesselation of the plane by equilateral triangles. A portion of such a tesselation is pictured in Figure 4.16.

Now, a subdivision of a face of the icosahedron is given by superimposing on this tesselation a larger equilateral triangle with lattice points for vertices. We have drawn the triangles for the subdivisions leading to I_2, I_3, I_4 and the subdivision just discussed (labeled J). A different subdivision, labeled K, is also included. In fact, all possible subdivisions can be constructed in the following way:

1. Pick any two lattice points and join them by an edge.
2. Rotate that edge 60 degrees about one of its endpoints to construct a new edge.
3. The new edge joins the center of rotation to another lattice point. (Why is this so?)
4. Fill in the last edge of the equilateral triangle.

Exercise 4.8. *Work out the details of the following classification scheme for geodesic domes due to H. S. M. Coxeter.*

Let C denote the collection (p, q) of the pairs of integers satisfying $0 \leq q$, $0 < p$. Superimpose a (p, q)-coordinate system on the tesselation in Figure 4.16 so that:

- *the origin is at a vertex of the tesselation;*
- *the positive p-axis is the horizontal ray of the tesselation from the origin and directed to the right;*
- *the positive q-axis is the ray of the tesselation from the origin making a 60-degree angle (measured counterclockwise) with the positive p-axis; and*
- *the sides of the triangles of the tesselation are one unit long.*

1. *Show that, for every* $(p, q) \in C$, *the points with coordinates* $(0, 0)$, (p, q) *and* $(-q, p + q)$ *are the vertices of an equilateral triangle.*
2. *Show that every equilateral triangle whose vertices are vertices of the tesselation is congruent to exactly one of the triangles described above.*
3. *Show that, in this classification scheme,* I_k *above corresponds to the pair* $(k, 0)$, *while* J *corresponds to* $(1, 1)$ *and* K *corresponds to* $(3, 1)$.
4. *Show that the equilateral triangle corresponding to* (p, q) *has area equal to* $p^2 + pq + q^2$ *times the area of a unit equilateral triangle.*
5. *Conclude that the geodesic dome corresponding to* $(p, q) \in C$ *has*
   (a) $10(p^2 + pq + q^2 - 1)$ *6-valent vertices;*
   (b) $10(p^2 + pq + q^2) + 2$ *vertices;*
   (c) $20(p^2 + pq + q^2)$ *faces;*
   (d) $30(p^2 + pq + q^2)$ *edges.*
6. *Verify that* $(7, 0)$ *and* $(5, 3)$ *correspond to the smallest geodesic domes of this type with the same number of vertices* (492) *but with different structure maps.*

In architecture, symmetry is a very pleasing and desirable property. Hence, there has been little interest in classifying *all* geodesic domes, including those in which the locations of the 5-valent vertices have no particular pattern. But, in its dual form, this classification problem is the focus of some very important and exciting research activity.

Recall how the dual of a map is constructed (at the end of Section 2.7), and note that the dual to the structure map of a geodesic dome is a trivalent (every vertex valence is 3) map with hexagonal and pentagonal faces. In fact, it has exactly 12 pentagonal faces. For example, the dual of the icosahedral

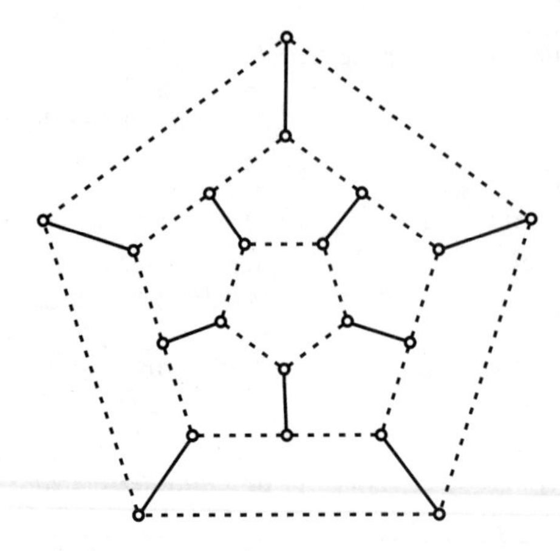

FIGURE 4.17

map is the dodecahedral map (pictured in Figure 4.17) while the dual to the graph given by subdivision J (pictured in Figure 4.15) is the soccer ball graph (pictured in Figure 4.18).

These are structure graphs for large molecules of carbon. Recall from Section 2.2 how graphs are used to represent chemical compounds. The dodecahedron and the soccer ball graphs are structure graphs of the carbon molecules C_{20} and C_{60}, respectively. The subscript denotes the number of carbon atoms in each molecule. Each carbon atom makes two single bonds and one double bond with nearby atoms. The convention is to represent these large carbon molecules by trivalent graphs without identifying the specific

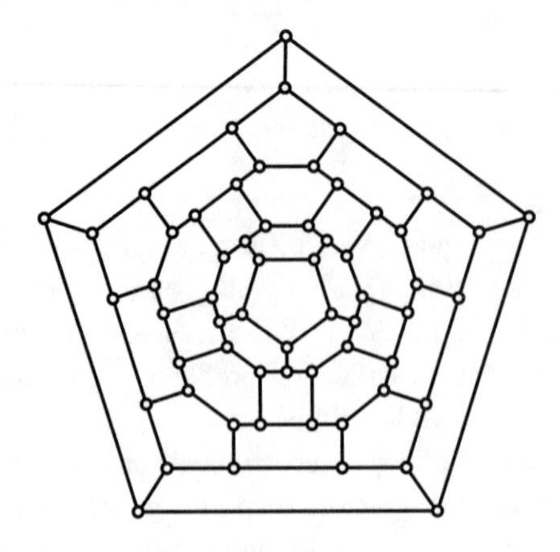

FIGURE 4.18

edges that correspond to the double bonds. In Figure 4.17, we have drawn the dodecahedron with one of the two possible ways (up to symmetry) of selecting edges to represent the double bonds—the double bonds are represented by solid lines and the single bonds by dashed lines.

Exercise 4.9. *Find the second way of selecting the edges of the dodecahedron to represent the double bonds in* C_{20} *and show that there are only two ways to do this up to symmetry.*

Large carbon atoms that have planar structure graphs with only hexagonal and pentagonal faces are called *fullerenes;* and the term has come to denote the class of all trivalent, plane graphs with only hexagonal and pentagonal faces. In the 1970s and 1980s several chemists (Osawa and others) predicted the stability of C_{60} with the soccer ball structure graph (Eiji Osawa, 1970). In the early 1980s astronomers (Walton, Oka and others) observed large carbon molecules, including C_{60}, streaming out of a red giant star. About the same time chemists (Rohlfing, Cox and Kalcor) were vaporizing graphite with a laser and observing a spike in the frequency tables at C_{60}.

The prevalence of C_{60} was taken as a sign that it was likely to be stable, and the soccer ball was the only trivalent graph on 60 vertices that, for chemical reasons, seemed to be stable. By the early 1990s enough evidence was amassed to convince the chemistry world that indeed a stable form of C_{60} with the soccer ball structure occurred and that many other fullerenes existed. The December 1991 issue of *SCIENCE* declared C_{60} the "molecule of the year"!

Chemists are interested in classifying *all* fullerenes and have set about doing so. *An Atlas of Fullerenes* lists and depicts most of hundreds of fullerenes of 100 vertices or fewer and gives estimates for the numbers of fullerenes with up to 140 vertices. But as yet no all-inclusive classification scheme for fullerenes is known.

In addition to simply classifying fullerenes, some chemists are trying to build them. Of particular interest are long tube-like fullerenes called *nanotubes. Scientific American* recently published two articles on the subject, "Tantalizing Tubes" and "Nanotubes: The Future of Electronics"; see the bibliography for more information. Also, several websites are devoted to nanotubes and possible applications. Searching on keywords like "nanotubes," "nanotechnology" and "nanoengineering" should lead to several such sites.

Returning to our subject of domes used to roof over a structure, it is natural to think of such a dome simply as a portion of a geodesic dome. In general, we have an embedded planar graph with an outer face Θ. The vertices on Θ represent points on the 3-dimensional framework that are fixed or pinned. We will use the degrees of freedom method for analyzing these structures. First, we make a simple observation:

- To triangulate a quadrilateral, we include one diagonal.
- Two diagonals are required to triangulate a pentagon.
- It takes three diagonals to triangulate a hexagon; and so on.

Since we will use this fact, you should prove it:

Exercise 4.10. *For all $n > 3$, prove the following two results.*

1. *Any planar n-gon can be triangulated by including some diagonals.*
2. *Any triangulation by diagonals of a planar n-gon uses exactly $n - 3$ diagonals.*

Let (V, E, F) be a planar graph embedded with all triangular faces save one, which is an n-gon. Adding $n - 3$ diagonals, we can make this graph a 3-tree. Hence,

$$3|V| - (|E| + n - 3) = 6$$

and the original graph has $3|V| - |E|$ or $3 + n$ degrees of freedom. To check that this fits with what we already know about such frameworks, we consider the first few values of n. If $n = 3$, that is, if the outer face is also triangular, then $3 + n = 6$ and the framework is rigid. Removing one nonredundant rod from a rigid framework results in a framework with 1 internal degree of freedom; removing one edge from a triangulated planar graph results in one quadrilateral face and 7 degrees of freedom (6 external and 1 internal). Taking $n = 4$ in our formula yields the same result.

Now, we wish to pin or fix the vertices of the n-gon and adjust the degrees of freedom count. Starting with $3 + n$ for the unpinned framework, we must subtract $3n$, the degrees of freedom of the vertices to be pinned, and add n for the edges of the n-gon that no longer reduce the degree of freedom. We get

$$3 + n - 3n + n = 3 - n.$$

So the pinned dome has $3 - n$ degrees of freedom. If $n = 3$, that is, if we fix the three vertices of a freestanding geodesic dome, the entire structure is fixed. If $n > 3$, the degree of freedom of the pinned framework is negative! We take this to mean that the pinned framework is overbraced and would remain rigid if up to $n - 3$ of the rods failed. In other words, a pinned framework with a planar structure graph may have some nontriangular faces and still be rigid. Before we consider frameworks with nontriangular faces, let's consider these (overbraced) domes with all faces triangular (except the rim).

Let (V, E, F) be a plane graph with all faces but one triangular. Assume that the outer face is an n-gon and assume that the typical vertex on the rim has valence 4 while the typical internal vertex has valence 6. Summing the vertex valences, we have

$$6|V| - 2n + d = 2|E|,$$

where d represents the sum of the deviations from the typical 4 or 6: a rim vertex of valence 3 or an internal vertex of valence 5 would contribute -1 to d, while a rim vertex of valence 6 or an internal vertex of valence 8 would contribute 2, and so on. We also have

$$3t + n = 2|E|,$$

where t is the number of triangular faces. And, of course, we have Euler's formula:

$$|V| - |E| + t + 1 = 2.$$

Exercise 4.11.

1. *Verify the above three equations.*
2. *Verify that they imply the following three equations.*
   (a) $d = -6$
   (b) $|E| = 3(|V| - 1) - n$
   (c) $t = 2(|V| - 1) - n$

The simplest example is one vertex of valence n with its neighbors forming the rim. For this graph, $|V| = n + 1$ and, as predicted, $|E| = 2n$ while $t = n$. The rim vertices all have valence 3, contributing $-n$ to d, while the only internal vertex has valence n, contributing $n - 6$ to d. Thus $d = -6$ as predicted.

Imagine cutting a geodesic dome along a geodesic; that is, imagine a circuit that passes only through vertices of valence 6 and having at each vertex two edges on the circuit and two on each side. The resulting "hemispheres" would have all 4-valent vertices on the rim and exactly 6 internal vertices of valence 5 (the rest being of valence 6). The number of vertices of valence 6 can

Courtesy Greg Vassallo

FIGURE 4.19

vary widely but depends on the placement of the vertices that contribute to d. For example, consider the domes over the water towers in Figure 4.11. A close-up view of one of the towers, with a 5-valent vertex highlighted, is given in Figure 4.19.

One can get nice views of the water towers from different sides, but since they are constructed on the highest hill in the area, one can only get partial views of the domes themselves. Furthermore, the closer one gets, the less of the domes one sees. Nevertheless, one can piece these fragments of information together to "reconstruct" the invisible parts of the domes.

- To start with, we will assume that the domes have a radial symmetry. This assumption is consistent with all views that one can get of the domes.
- Second, we note that several 5-valent vertices occur in the first row of vertices above the base and that no 5-valent vertices are be seen at any other level. One possibility for another 5-valent vertex would be at the top of the dome. This is a very reasonable assumption, since such a vertex would make it easier to accommodate the curvature of the dome with minimal deformation of the triangles.
- We can estimate the number of edges in the base by directly counting around the towers and estimating the few in the region of the tangency of the towers. Doing this we will conclude that, for each of the smaller towers, the length of its rim is between 23 and 27, while the large central tower's rim length is in the range of 35 to 40.

- Finally, walking around the base reveals that each vertex on the base has valence 4. See Figure 4.19.

These assumptions now force the entire structure of the smaller domes. The assumptions of a central 5-valent vertex and radial symmetry force the remaining 5-valent vertices to be evenly spaced and the base circuit to have length a multiple of 5, that is, 25.

Since each of the 25 vertices of the base has valence 4, there are also 25 vertices in the second ring of vertices. But, the five 5-valent vertices in this second ring force each successive ring of vertices to be shorter by 5: ring 3 has length 20; ring 4, length 15; ring 5, length 10; and ring 6, length 5. There are

$$25 + 20 + 15 + 10 + 5 = 75$$

edges on these inner rings, and each is the base of two triangles. So adding the 25 triangles with base on the rim gives

$$t = 2 \times 75 + 25 = 175$$

triangles. From this we conclude that $|V| = 101$ and $|E| = 275$.

Exercise 4.12. *Verify the above computations.*

Exercise 4.13. *Actually, before the leaves fell from the trees, it was impossible to get an accurate count of the spacing of the* 5*-valent vertices in the second rim. We concluded that the rim had length* 24*, that the* 5*-valent vertices occurred in every sixth position on the second rim and that there was a single* 4*-valent vertex at the top. Deduce the structure of such a dome.*

Exercise 4.14. *Based on the information given above, deduce the possible structures for the central dome.*

Exercise 4.15. *Consider covering the seats of a stadium by a dome with its center removed. The structure graph is a triangulated annulus with inner rim of length m and outer rim of length n. Assume as usual that the outer rim is pinned. Compute the degrees of freedom of such a covering. Assume k-fold rotational symmetry; that all outer rim valences are* 4*; that all inner rim vertices have valences* 4 *and* 5*; that all internal* 5*-valent vertices are in the first tier above the outer rim; and that all remaining vertices have valence* 6*. Discuss the possible lengths for m and the number of triangular faces.*

4.4 Tensegrity: Tension Bracings

The very first problem we considered in detail was the grid bracing problem. We reconsider that problem now with one rather natural alteration. Consider the braced 1×1-grid in Figure 4.20. If the rigid brace were to be replaced by a wire or cable, the cell could be deformed—but only in one direction. The cell could then be made rigid by including a second wire along the other diagonal. A bracing of the $m \times n$-grid using diagonal wires to brace some of the cells is called a *tension bracing.*

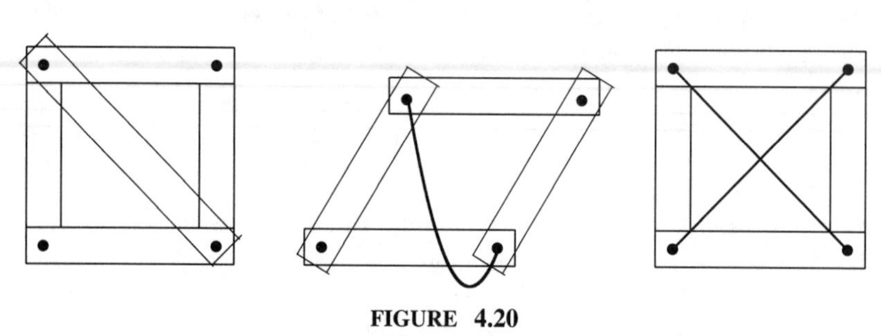

FIGURE 4.20

One way to get a rigid tension bracing of a grid is to first find an ordinary bracing that makes it rigid and then replace each brace by two tension wires. Of course, these are probably not the only tension bracings for most grids. One might expect many bracings in which some cells have only wires in the lower left to upper right direction while others are braced in the other direction. The $1 \times k$-grid and the $k \times 1$-grid are exceptions in that these grids need both wires in all cells to be rigid. This follows from Lemma 4.8, which you should have no trouble proving. To simplify our discussion, let's call a wire from the lower left corner of a cell to its upper right corner an NE (for northeast) wire or cable; one in the other direction is an NW (for northwest) wire or cable.

Lemma 4.8. *A rigid tension bracing of the $m \times n$-grid must include both an NE cable and an NW cable in each row and column.*

Exercise 4.1. *Prove Lemma 4.8.*

The next simplest case to consider is the 2×2-grid. Here an ordinary rigid bracing requires rigid braces in three of the four cells. Replacing each of these by a pair of wires gives a rigid tension bracing with three NE and three NW wires. The interesting fact is that, while all six of these braces are necessary in this bracing, one can make the grid rigid with only four wires!

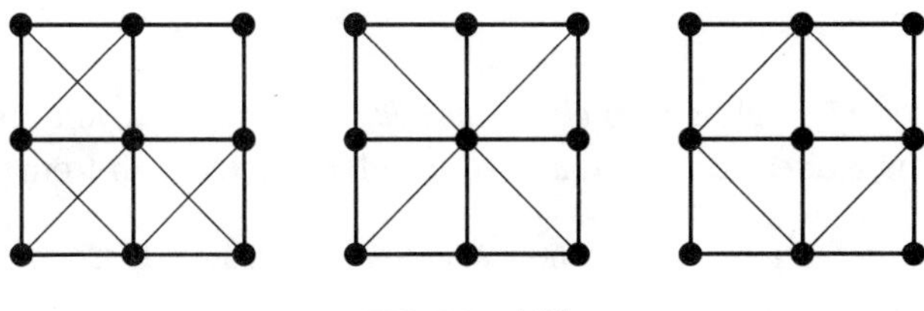

FIGURE 4.21

Exercise 4.2. *Consider the motions of the unbraced* 2×2*-grid and then consider the tension bracings pictured in Figure 4.21.*

1. *Show that each of these bracings is rigid.*
2. *Show that each is minimal in that the removal of any wire from any of these bracings results in a nonrigid tension bracing.*

The results in Exercise 4.2 are surprising. But only after some thought do we realize that these examples also show something very significant:

> *There is no natural way to define degrees of freedom for tension bracings.*

What could the degrees of freedom be for the unbraced 2×2-grid? Four? Six?

Even though the degrees of freedom approach does not seem to work, a variation on the bipartite graph model will. The question is how to code the directions of the bracing wires in the associated bipartite graph. To incorporate this additional information, we employ a *directed graph.* We may think of a directed graph as an ordinary graph with directions assigned to its edges. Each edge can have either of two directions, and directed graphs permit edges in both directions between a pair of vertices.

In Figure 4.22, we illustrate the coding of a tension bracing with a bipartite, directed graph. To do this we arbitrarily decide that an NE edge will be represented by an edge directed upward and an NW edge by a downward directed edge.

The reader wishing to pursue this model will, of course, have to develop some of the basics of the theory of directed graphs. The important definitions are that of a *directed path* and a *strongly connected* directed graph:

- A *directed path* from x to y is a path $x = x_0, x_1, x_2, \ldots, x_k = y$ so that, for $i = 0$ to $n - 1$, the edge joining x_i and x_{i+1} is directed from x_i to

x_{i+1}.

- A directed graph is *strongly connected* if, for any two vertices x and y, there is a directed path from x to y and a directed path from y to x.

With these definitions, we can now state the analogue to Theorem 2.22:

Theorem 4.9. *A tension braced grid will be rigid if and only if its associated bipartite directed graph is strongly connected.*

This graph model permits us to explain the failure of the degrees of freedom approach to tension bracings. Theorem 2.22 states that an (ordinary) braced grid will be rigid if and only if its associated bipartite graph is connected. So rigid bracings are based on connectivity. The (internal) degree of freedom of a subgraph is the minimal number of edges one must add to make a connected spanning subgraph. This definition works because, no matter what algorithm you may use to build up a spanning connected subgraph, it will result in a spanning tree, and all spanning trees have the same number of edges. In short, all *minimal* connected spanning subgraphs are *minimum* connected spanning subgraphs.

The corresponding statement is not true for strong connectivity in directed graphs. Minimal, strongly connected, spanning subgraphs of a directed graph may be of differing sizes. Hence the number of directed edges you need to add to a subgraph to make it strongly connected could vary depending on how you choose to add edges.

This discussion is illuminated by the next exercise.

Exercise 4.3. *Consider the* 2×2*-grid as pictured in Figure 4.21.*

1. *Sketch the complete, directed, bipartite graph corresponding to the grid with all wire braces included.*

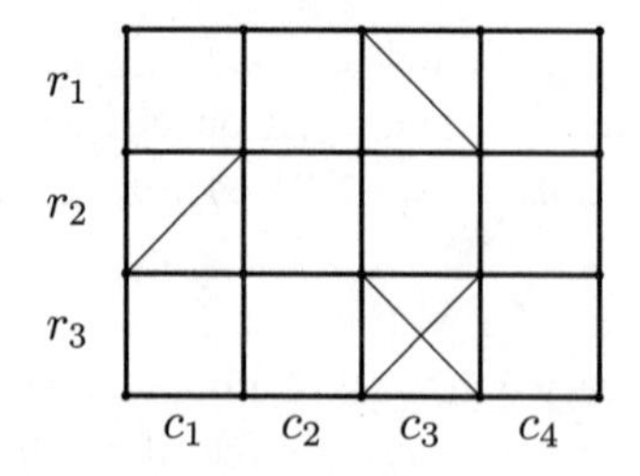

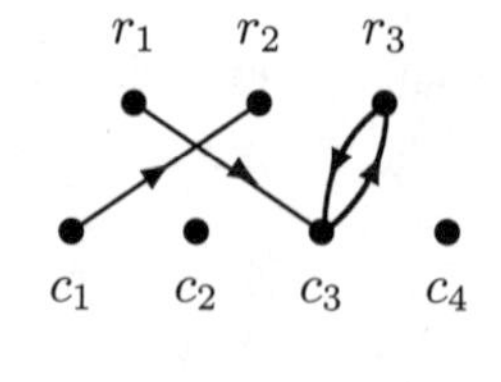

FIGURE 4.22

2. *Show that the subgraphs corresponding to each of the tension bracings in Figure 4.21 are minimal, strongly connected, spanning subgraphs.*

3. *Show that the minimum, strongly connected, spanning subgraphs have four edges and that there are only two of them.*

4. *Consider the six-wire minimal bracing pictured on the left in Figure 4.21. Add the wires one at a time, and observe that at each stage the collection of motions is reduced until only the rigid motions are left.*

Exercise 4.4. *Use Theorem 4.9 to decide on the minimum number of wire braces that are needed to produce a rigid tension bracing of the* $2 \times k$*-grid.*

Tension bracings of grids set the tone for the entire discussion of tension frameworks or *tensegrity frameworks*, as they are usually called. The combinatorial, degrees of freedom approach no longer works. The tools that we need to study tension frameworks are an expansion of those we developed for investigating infinitesimal rigidity. As the reader may suspect, frameworks consisting of rods and cables held in place by tension are more complicated than the frameworks that we have discussed so far. In fact, the theory of tensegrity frameworks is not as well understood. Hence, we propose to move into a much more intuitive mode, starting with a rather intuitive definition of *tensegrity framework:*

- By an *m-dimensional, rigid, tensegrity framework* we mean a framework in m-space consisting of cables and rods with the cables being stretched and the rods being compressed such that the various parts of the framework do not move relative to one another.

Perhaps the easiest way to construct an m-dimensional, rigid, tensegrity framework is to start with an m*-circuit,* that is, an m-dimensional framework with the property that the removal of any rod leaves an m-tree. Now remove any rod from this framework; the resulting framework is still rigid. Replace the removed rod by a cable with a turnbuckle (a tightening device), and tighten the cable until it is under stress. We claim that every other rod in the framework will be put under stress: Some will be under a compression stress, others under a stretching stress. Assuming this to be true, replace those rods that are under a stretching stress by cables. The result is a tensegrity framework.

To verify our claim, let r_0 denote the rod to be replaced by the turnbuckle and let r denote any other rod. If we remove both rods, the resulting framework is no longer rigid, so it admits deformations. But note that any deformation

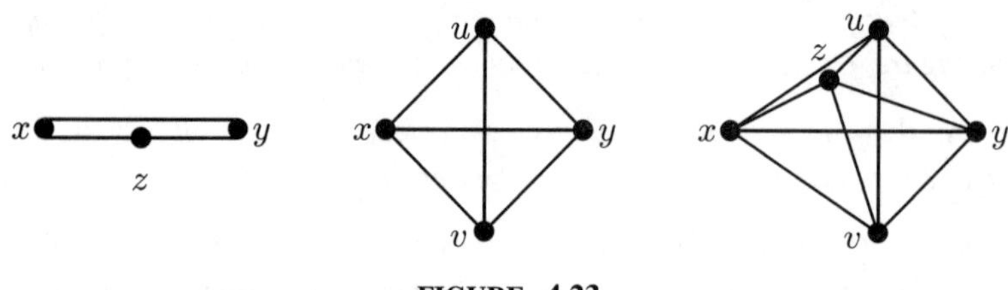

FIGURE 4.23

must change the distance between the endpoints of both of the missing rods—any deformation that did not change the distance between the endpoints of r_0 (r) would be a deformation of the framework with r_0 (r) included. Thus, tightening the turnbuckle with r removed would either increase or decrease the distance between the endpoints of r; tightening the turnbuckle with r included would either stretch or compress r.

Let's illustrate this construction using the simplest m-circuits in dimensions 1, 2 and 3. They are pictured in Figure 4.23.

Consider the 1-dimensional framework from Figure 4.23. If we replace the x, y-rod by a cable under tension, the remaining two rods are compressed, giving one of the simplest 1-dimensional tensegrity frameworks (in the next exercise, you will construct the other). If we replace the x, y-rod in the 2-dimensional framework by a cable under tension, we can see that the four rods (x, u), (x, v), (y, u) and (y, v) will be compressed, while the remaining rod, between u and v, will be stretched. Replacing both the x, y-rod and the u, v-rod by a wire results in one of the simplest 2-dimensional tensegrity frameworks.

As we move on to the 3-dimensional case, it becomes more difficult to simply talk our way through what will happen when one rod is replaced by a cable under tension. First, we must make sure that we are all interpreting the 2-dimensional representation of the right-hand 3-dimensional framework in the same way:

- Visualize vertices x and y as being in the plane of the page.
- Visualize vertex z as being in front of the plane of the page.
- Visualize vertices u and v as being slightly behind the page so the x, y-edge is perpendicular to the u, v, z-triangle and passes through its interior.

With this understanding, the reader may be able to visualize the tensegrity framework with cables under a stretching stress in place of the x, y-edge and the three edges of the u, v, z-triangle; the remaining six rods will then be under compression stress. It is clear that very soon the frameworks will

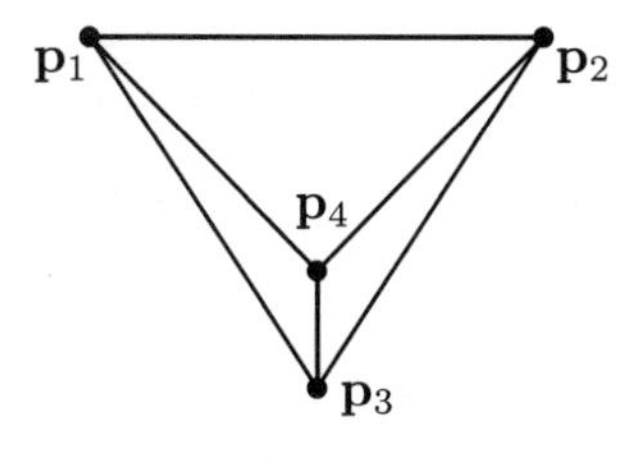

FIGURE 4.24

get too large for this kind of analysis. So our next task is to develop some mathematical tools for designing tensegrity frameworks.

Exercise 4.5. *For the three tensegrity frameworks just described, interchange the roles of the edges: Replace all cables by rods and all rods by cables. Convince yourself that, in each case, the resulting tensegrity framework is also rigid.*

Consider the 2-dimensional framework $\mathcal{F}$ pictured in Figure 4.24. To talk our way through this one, we might do the following:

- Delete the rods joining $\mathbf{p}_1$ to $\mathbf{p}_2$ and $\mathbf{p}_3$ to $\mathbf{p}_4$.
- Move $\mathbf{p}_1$ and $\mathbf{p}_2$ closer together.
- Note that this forces both $\mathbf{p}_3$ and $\mathbf{p}_4$ to move downward.

Now, if $\mathbf{p}_3$ is moving faster than $\mathbf{p}_4$, a tension wire from $\mathbf{p}_1$ to $\mathbf{p}_4$ would stretch the $\mathbf{p}_3\mathbf{p}_4$ rod and put the remaining four rods under compression. But if $\mathbf{p}_4$ is moving faster than $\mathbf{p}_3$, a tension wire from $\mathbf{p}_1$ to $\mathbf{p}_2$ would compress the $\mathbf{p}_3\mathbf{p}_4$, $\mathbf{p}_1\mathbf{p}_4$ and $\mathbf{p}_2\mathbf{p}_4$ rods while stretching the $\mathbf{p}_1\mathbf{p}_3$ and $\mathbf{p}_2\mathbf{p}_3$ rods. Which is it?

To answer this question, suppose that the $\mathbf{p}_1\mathbf{p}_2$ rod is replaced by a wire under tension. Now all rods of the framework will be under stress and these stresses can be represented by forces along the rods. Furthermore, these forces must be in equilibrium at the vertices, giving rise to a system of vector equations that must be satisfied.

- The force at $\mathbf{p}_1$ due to the stress in the wire is represented by a vector at $\mathbf{p}_1$ directed toward $\mathbf{p}_2$ or a scalar multiple of $\mathbf{p}_2 - \mathbf{p}_1$. It will be convenient to denote this vector by $s_{12}\,(\mathbf{p}_1 - \mathbf{p}_2)$ where s_{12} is negative.
- The force at $\mathbf{p}_2$ due to the stress in the wire is then $s_{12}\,(\mathbf{p}_2 - \mathbf{p}_1)$.
- The forces due to the stresses in the other rods at $\mathbf{p}_1$ are $s_{14}\,(\mathbf{p}_1 - \mathbf{p}_4)$ and $s_{13}\,(\mathbf{p}_1 - \mathbf{p}_3)$; the scalars will be negative if the rod is being stretched and positive if is is being compressed.

Also, these three vectors must be in equilibrium or the joint $\mathbf{p}_1$ would move. So we have

$$s_{12}(\mathbf{p}_1 - \mathbf{p}_2) + s_{14}(\mathbf{p}_1 - \mathbf{p}_4) + s_{13}(\mathbf{p}_1 - \mathbf{p}_3) = 0.$$

Similarly,

$$s_{12}(\mathbf{p}_2 - \mathbf{p}_1) + s_{24}(\mathbf{p}_2 - \mathbf{p}_4) + s_{23}(\mathbf{p}_2 - \mathbf{p}_3) = 0;$$

$$s_{13}(\mathbf{p}_3 - \mathbf{p}_1) + s_{34}(\mathbf{p}_3 - \mathbf{p}_4) + s_{23}(\mathbf{p}_3 - \mathbf{p}_2) = 0;$$

$$s_{14}(\mathbf{p}_4 - \mathbf{p}_1) + s_{34}(\mathbf{p}_4 - \mathbf{p}_3) + s_{24}(\mathbf{p}_4 - \mathbf{p}_2) = 0.$$

If we assign coordinate to the points, we may solve for the stresses. Take $\mathbf{p}_1 = (-2,3)$, $\mathbf{p}_2 = (2,3)$, $\mathbf{p}_3 = (0,0)$ and $\mathbf{p}_4 = (0,1)$. This system of vector equations then becomes

$$\begin{array}{llllllll}
s_{12}(-4,0) & +s_{13}(-2,3) & +s_{14}(-2,2) & & & & = (0,0) \\
s_{12}(4,0) & & & +s_{23}(2,3) & +s_{24}(2,2) & & = (0,0) \\
& s_{13}(2,-3) & & +s_{23}(-2,-3) & & +s_{34}(0,-1) & = (0,0) \\
& & s_{14}(2,-2) & & +s_{24}(-2,-2) & +s_{34}(0,1) & = (0,0)
\end{array}$$

This is equivalent to following system of equations:

$$\begin{array}{rrrrrrl}
-4s_{12} & -2s_{13} & -2s_{14} & & & & =0 \\
& +3s_{13} & +2s_{14} & & & & =0 \\
4s_{12} & & & +2s_{23} & +2s_{24} & & =0 \\
& & & +3s_{23} & +2s_{24} & & =0 \\
& 2s_{13} & & -2s_{23} & & & =0 \\
& -3s_{13} & & -3s_{23} & & -s_{34} & =0 \\
& & 2s_{14} & & -2s_{24} & & =0 \\
& & -2s_{14} & & -2s_{24} & +s_{34} & =0
\end{array}$$

Solving this redundant system yields

$$s_{13} = s_{23} = 4s_{12},$$

$$s_{14} = s_{24} = -6s_{12},$$

$$s_{34} = -24s_{12}.$$

Thus, if the $\mathbf{p}_1\mathbf{p}_2$ rod is replaced by a tension cable, s_{12}, s_{13} and s_{23} are negative while s_{14}, s_{24} and s_{34} are positive. Specifically, we may take s_{12} to be -1, then $s_{13} = s_{23} = -4$, $s_{14} = s_{24} = 6$ and $s_{34} = 24$. These numbers are called *stresses*. The collection of stresses is called *a collection of resolvable stresses* for the framework $\mathcal{F}$.

Given a set of resolvable stresses for a (rod and joint) framework, one may then construct a tensegrity framework by replacing all rods assigned negative stresses by cables under tension; the remaining rods will then be compressed. The tensegrity framework for $\mathcal{F}$ given by the set of stresses just computed consists of three (compression) rods emanating from $\mathbf{p}_4$ and three (tension) cables around the outside. Taking s_{12} to be positive, on the other hand, changes all signs and interchanges the roles of cables and rods.

Exercise 4.6. *Referring to Figure 4.24, take* $\mathbf{p}_1 = (-2,3)$, $\mathbf{p}_2 = (2,3)$, $\mathbf{p}_3 = (0,0)$ *and* $\mathbf{p}_4 = (0,y)$*; that is, let* $\mathbf{p}_4$ *take any position on the* y*-axis.*

1. *Solve the system in terms of* s_{12} *and* y.
2. *Describe the configurations of cables and rods in each of the cases:* $y < 0$, $0 < y < 3$ *and* $y > 3$.

Consider the framework in Figure 4.24 as a "standard" 2-dimension rod and joint framework, and write down its rigidity matrix in vector form:

$$\mathbf{M}(\mathcal{F}) = \begin{bmatrix} \mathbf{p}_1 - \mathbf{p}_2 & \mathbf{p}_2 - \mathbf{p}_1 & & \\ \mathbf{p}_1 - \mathbf{p}_3 & & \mathbf{p}_3 - \mathbf{p}_1 & \\ \mathbf{p}_1 - \mathbf{p}_4 & & & \mathbf{p}_4 - \mathbf{p}_1 \\ & \mathbf{p}_2 - \mathbf{p}_3 & \mathbf{p}_3 - \mathbf{p}_2 & \\ & \mathbf{p}_2 - \mathbf{p}_4 & & \mathbf{p}_4 - \mathbf{p}_2 \\ & & \mathbf{p}_3 - \mathbf{p}_4 & \mathbf{p}_4 - \mathbf{p}_3 \end{bmatrix}$$

To get the rigidity matrix as we first defined it, we simply replace each column of vectors by the two columns of their coordinates. Thus $\mathbf{M}(\mathcal{F})$ equals

$$\begin{bmatrix} (x_1-x_2) & (y_1-y_2) & (x_2-x_1) & (y_2-y_1) & 0 & 0 & 0 & 0 \\ (x_1-x_3) & (y_1-y_3) & 0 & 0 & (x_3-x_1) & (y_3-y_1) & 0 & 0 \\ (x_1-x_4) & (y_1-y_4) & 0 & 0 & 0 & 0 & (x_4-x_1) & (y_4-y_1) \\ 0 & 0 & (x_2-x_3) & (y_2-y_3) & (x_3-x_2) & (y_3-y_2) & 0 & 0 \\ 0 & 0 & (x_2-x_4) & (y_2-y_4) & 0 & 0 & (x_4-x_2) & (y_4-y_2) \\ 0 & 0 & 0 & 0 & (x_3-x_4) & (y_3-y_4) & (x_4-x_3) & (y_4-y_3) \end{bmatrix}$$

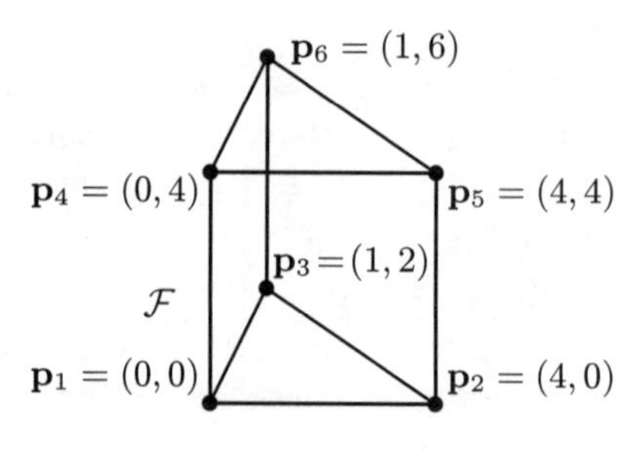

FIGURE 4.25

Considering the vector form of the rigidity matrix for $\mathcal{F}$, we see that a set of resolvable stresses for $\mathcal{F}$ is nothing more than the set of coefficients of a relation among the rows of $\mathbf{M}(\mathcal{F})$! Thus, it is not necessary to start with a birigid framework; one simply needs a framework with the property that the rows of its rigidity matrix are not independent. Hence, another source of tensegrity frameworks is generically isostatic frameworks that are not generically embedded. In Figure 4.25, we picture such a framework—one that we discussed earlier.

Recall that the framework $\mathcal{F}$ is not rigid nor infinitesimally rigid: The top triangle can move to the right and down simultaneously, distorting the three parallelograms connecting it with the bottom triangle. In Exercise 3.2, you verified that the following matrix (with its rows labeled by the corresponding rods) is the rigidity matrix for this framework:

$$\begin{bmatrix}
0 & -4 & 0 & 0 & 0 & 0 & 0 & 4 & 0 & 0 & 0 & 0 \\
0 & 0 & 0 & 0 & 0 & 0 & -4 & 0 & 4 & 0 & 0 & 0 \\
0 & 0 & 0 & 0 & 0 & 0 & -1 & -2 & 0 & 0 & 1 & 2 \\
0 & 0 & 0 & 0 & 0 & 0 & 0 & 0 & 3 & -2 & -3 & 2 \\
0 & 0 & 0 & -4 & 0 & 0 & 0 & 0 & 0 & 4 & 0 & 0 \\
-4 & 0 & 4 & 0 & 0 & 0 & 0 & 0 & 0 & 0 & 0 & 0 \\
-1 & -2 & 0 & 0 & 1 & 2 & 0 & 0 & 0 & 0 & 0 & 0 \\
0 & 0 & 3 & -2 & -3 & 2 & 0 & 0 & 0 & 0 & 0 & 0 \\
0 & 0 & 0 & 0 & 0 & -4 & 0 & 0 & 0 & 0 & 0 & 4
\end{bmatrix}
\begin{matrix}
(\mathbf{p}_1 - \mathbf{p}_4) \\
(\mathbf{p}_4 - \mathbf{p}_5) \\
(\mathbf{p}_4 - \mathbf{p}_6) \\
(\mathbf{p}_5 - \mathbf{p}_6) \\
(\mathbf{p}_2 - \mathbf{p}_5) \\
(\mathbf{p}_1 - \mathbf{p}_2) \\
(\mathbf{p}_1 - \mathbf{p}_3) \\
(\mathbf{p}_2 - \mathbf{p}_3) \\
(\mathbf{p}_3 - \mathbf{p}_6)
\end{matrix}$$

You also verified in Exercise 3.2, that the framework has rank 8; that is, its rows are dependent. Thus, this framework admits a set of resolvable stresses. It is not difficult to verify that the scalars listed here are such a set:

$$\begin{aligned}
&s_{14} = 6, \quad s_{45} = -3, \quad s_{46} = 12, \quad s_{56} = 4, \quad s_{25} = 2, \\
&s_{12} = 3, \quad s_{13} = -13, \quad s_{23} = -4, \quad s_{36} = -8.
\end{aligned}$$

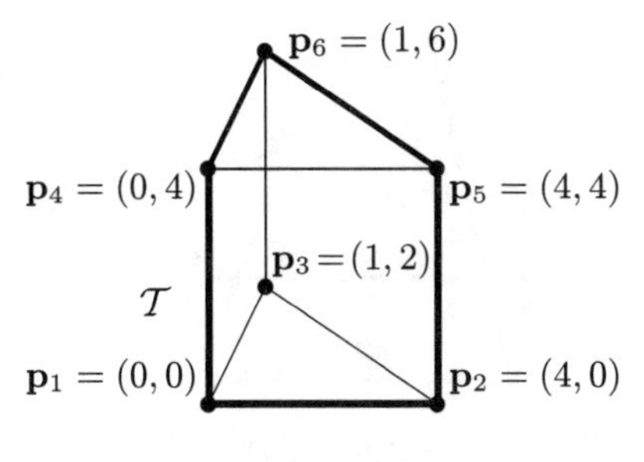

FIGURE 4.26

The corresponding tensegrity framework $\mathcal{T}$ is pictured in Figure 4.26, where the thick lines represent rods and the thin lines represent the tension wires.

Exercise 4.7. *Verify that the set of scalars listed above is a set of resolvable stresses for $\mathcal{F}$.*

We were able to construct tensegrity framework $\mathcal{T}$ from $\mathcal{F}$ because $\mathcal{F}$ admitted a nontrivial motion. The natural question to ask is:

Does $\mathcal{T}$ admit the same motion?

This question is not so easy to answer. First let's suppose that you constructed a model of $\mathcal{T}$ in which the wires were the exact length but not under any stress. It is pretty clear that this unstressed model should admit the motion.

Now suppose that you had a model of $\mathcal{T}$ with the set of resolvable stresses computed above. If you were to apply the motion to this model it would move. But once you let it go, it would move back to the original position!

Why?

The motion of $\mathcal{F}$ does not change any of the lengths of the rods or wires—not even infinitesimally. Hence the motion cannot change the stresses in the individual wires and rods. Thus, in the new position, the stresses are not in equilibrium at the joints, and the joints move until equilibrium is restored. We have arrived at a rather paradoxical observation:

A nonrigid framework can be made rigid simply by putting its members under appropriate stresses.

Actually, this example illustrates the fact that there are two levels of "rigidity" for tensegrity frameworks. The tensegrity frameworks based on the frameworks in Figures 4.23 and 4.24 cannot be deformed without stretching a cable or bending a rod. On the other hand, the framework just discussed resists

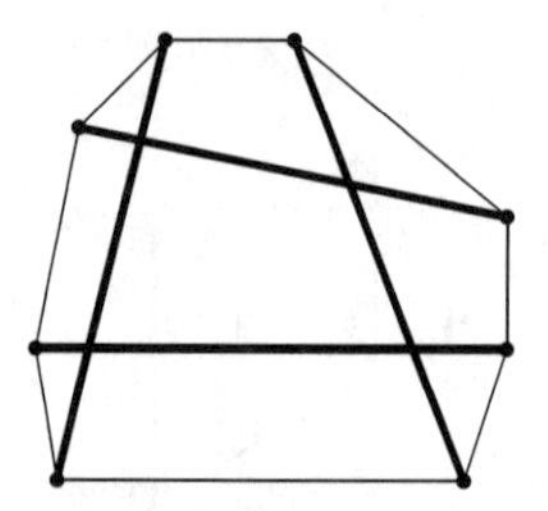

FIGURE 4.27

being deformed but, under an outside force, will deform without changing the lengths of any of its cables or rods.

Even more striking examples of this latter type can be constructed using planar rod and joint frameworks that have n joints but fewer than $2n-3$ rods. Such a tensegrity framework is pictured in Figure 4.27; again, the thick lines represent rods and the thin lines represent the tension wires. The structure graph for this framework has 8 vertices and 12 edges—one edge short of the 13 ($2 \times 8 - 3$) needed for rigidity. Nevertheless, as explained above, the tensegrity framework in Figure 4.27 (with the appropriate stresses) will resist any motion. In fact, such frameworks are very easy to construct: Simply put the framework together from rods and elastic cables, and the framework then pulls itself into equilibrium!

How far can we go? Can any 2-dimensional framework be converted into a rigid framework? The answer is no, as we now show.

Let (V, E) be the structure graph of a rigid tensegrity framework $\mathcal{T}$ in the plane. We can make a few simple observations: Clearly, (V, E) must be connected, and it may not have any pendant vertices. What about vertices of valence 2? Suppose (V, E) has a vertex of valence 2 and $\mathbf{p}$ is the corresponding joint of $\mathcal{T}$. Since the stress vectors of the two members at $\mathbf{p}$ must add to the zero vector, these vectors must be equal in magnitude and opposite in direction. There are several possibilities:

- Two rods could meet at $\mathbf{p}$. In this case, the angle at $\mathbf{p}$ is a straight angle. Thus, the two rods could be replaced by a single rod as pictured in Figure 4.28(a).
- Two cables could meet at $\mathbf{p}$. In this case, the angle at $\mathbf{p}$ is also a straight angle. Thus, the two cables could be replaced by a single cable as pictured in Figure 4.28(b).
- A rod and cable could meet at $\mathbf{p}$. In this case, the angle at $\mathbf{p}$ is zero. If the rod is longer than the cable, the pair could be replaced by a single

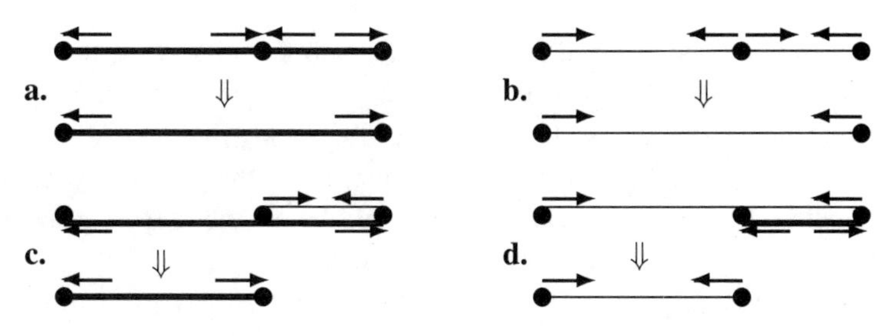

FIGURE 4.28

rod as pictured in Figure 4.28(c); and, if the rod is shorter than the cable, the pair could be replaced by a cable as pictured in Figure 4.28(d).

- There is one other possibility: We could have two rods meeting at $\mathbf{p}$ at some angle other than 180 degrees. But, in this case, the stresses in both rods must be zero and $\mathcal{T}$ must consist a smaller rigid tensegrity framework $\mathcal{T}'$ to which the two rods have been attached. In this case, we may simply delete the two rods.

It follows that any rigid tensegrity framework with vertices of valence 2 can be reduced to a smaller rigid tensegrity framework without vertices of valence 2. Henceforth we will consider only tensegrity frameworks with structure graphs that have vertex valences 3 or more.

We should note that all of the configurations in Figure 4.28 except configuration (b) are unstable. For example, the slightest push up or down on the center joint of the configuration (a). would cause it to buckle. We also see from this discussion that the cable–rod symmetry is not perfect: Configuration (b) is stable, but interchanging cables and rods yields configuration (a), which is not stable.

The condition that every vertex of the graph (V, E) has valence at least 3 implies that $|E| \geq \frac{3}{2}|V|$. Compare this with the condition required if (V, E) is to be generically 2-rigid: $|E| \geq 2|V| - 3$. So it seems that a tensegrity framework could be rigid with far fewer rods and cables than the number of rods necessary for the corresponding ordinary framework to be rigid. Indeed, we can verify that rigid tensegrity frameworks exist with $|V|$ arbitrarily large and $|E| = \frac{3}{2}|V|$.

First note that $|E| = \frac{3}{2}|V|$ implies that $|V|$ is even. Let $|V| = 2n$ and let (V, E) consist of a circuit of length $2n$ with opposite vertices (around the circuit) joined by an edge. One easily checks that this is the structure graph of a rigid tensegrity framework $\mathcal{T}$. Take the $2n$ joints to be evenly spaced about a circle of radius 1 and joined by cables to form a circuit; then insert a rod between opposite joints. If a stress of -1 is applied to the cables, a counterbalancing stress of $2\sin^2(180/n)$ in the rods will result in the rigid tensegrity framework we seek. In fact, as n gets large, there are many ways in which a circuit of $2n$ cables can be made rigid by adding n rods; of course, the cables and rods can also be of varying lengths.

These observations are illustrated by the rigid tensegrity framework in Figure 4.27.

Exercise 4.8. *Computing the actual stresses in an arbitrary tensegrity framework can be quite tedious. In general, the only tractable cases with large n are highly symmetric.*

1. *Prove that the stresses -1 in the cables and $2\sin^2(180/n)$ in the rods is a resolvable set of stresses for the tensegrity framework described above. Refer to the left-hand picture in Figure 4.29 and make the following computations:*

   (a) *Show that the measure of angle α is $180/n$ degrees.*

   (b) *Show that the length of each cable is $2\sin(\alpha)$ units.*

   (c) *Show that the magnitude of the stress vectors in the cable (under a stress factor of -1) is then $2\sin(\alpha)$.*

   (d) *Show that the projection of the cable stress vector onto the rod has magnitude $2\sin^2(\alpha)$.*

   (e) *Conclude that the stress factor for the rods is $2\sin^2(\alpha)$.*

2. *Find a resolvable set of stresses for the other tensegrity frameworks in Figure 4.29. In both cases, assume that the joints are evenly spaced around a circle of radius 1.*

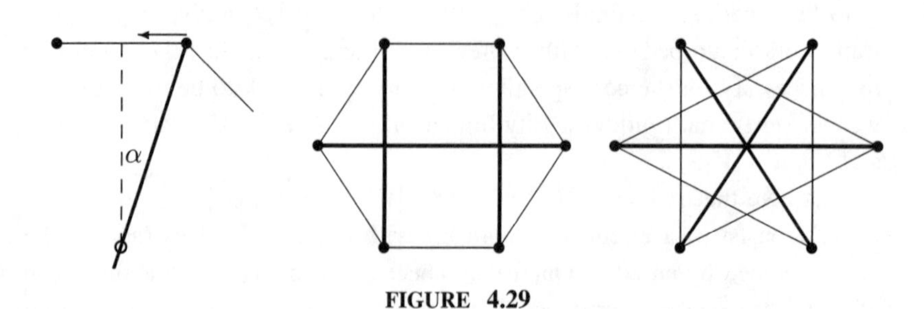

FIGURE 4.29

Photo by Jerry L. Thompson.

FIGURE 4.30
Free Ride Home, 1974, Kenneth Snelson, Storm King Art Center

Exercise 4.9. *The ratio of cables to rods in the above examples is* 2 *to* 1. *The fewer the rods, the more striking the tensegrity framework is—particularly for the* 3*-dimensional tensegrity frameworks. Hence it is natural to ask just how large the ratio of cables to rods can be. It turns out that this ratio can be as large as you wish. Verify this.*

Of course, the most interesting tensegrity frameworks are 3-dimensional. The space sculptures of Kenneth Snelson seem to defy gravity. Figure 4.30 is a photograph of his "Free Ride Home" at the Storm King Art Center.

It has even been suggested that we ourselves are 3-dimensional rod (bone) and cable (tendon, muscle, ligament) tensegrity frameworks! In a January 1998 *Scientific American* article, Donald Ingber discusses "The Architecture of Life," tracing the role of tensegrity in living organisms at all levels from the macro down to the cellular. Shortly afterward, in the March–April 1998 issue of *American Scientist*, Robert Connelly and Allen Back gave an excellent introduction to tensegrity in their article "Mathematics and Tensegrity."

The simplest 3-dimensional tensegrity framework is based on the simplest 3-circuit, K_5. We discussed this framework at the beginning of this section; we take a closer look now. Two fundamentally different tensegrity frameworks

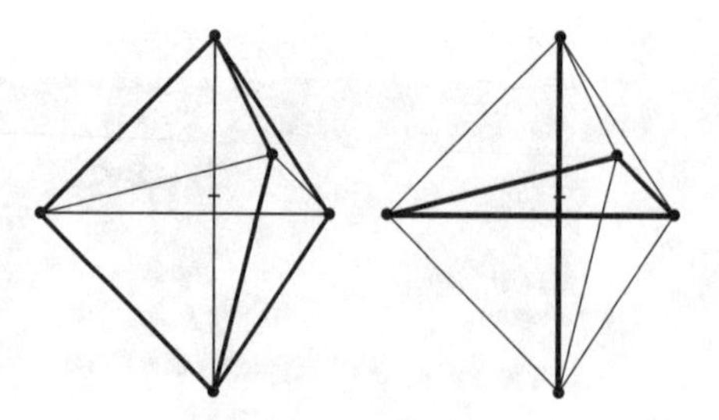

FIGURE 4.31

can be based on it.

- The first variant of the simplest 3-dimensional integrity framework is constructed by embedding the five vertices so that no one lies in the tetrahedron formed by the other four.

In this case, the ordinary framework will consist of two tetrahedra joined along a common face forming a convex polyhedron with the two vertices opposite the common face joined by a segment through the interior of the polyhedron.

Consider the left-hand framework in Figure 4.31; the top and bottom points are vertices of the two tetrahedra having the middle triangle as their common face. The segment joining the top and bottom points pierces the plane of the middle triangle at the tick mark. If this interior segment is replaced by a cable under stress, the three rods of the common tetrahedral face will be stretched and may also be replaced by cables. This tensegrity framework is pictured on the left in Figure 4.31; its complementary tensegrity framework is on the right.

- The second pair of tensegrity frameworks based on K_5 come by embedding four of the vertices as the joints of a tetrahedron and placing the fifth joint in its interior.

First, take the tetrahedron made from rods and connect the fifth joint to each other joint by a cable under tension. Interchanging the roles of cable and rod gives the "complementary" tensegrity framework consisting of a wire tetrahedron held under tension by four rods from an interior joint. Again the symmetry is not perfect: In the first case, the tetrahedron of rods is rigid on its own, but no part of the second framework can stand alone.

Tensegrity frameworks that are 3-dimensional are easy to construct. One can use dowels, string and elastic bands, even soda straws to build interesting models. The two models in Figure 4.32 were constructed using a commercial

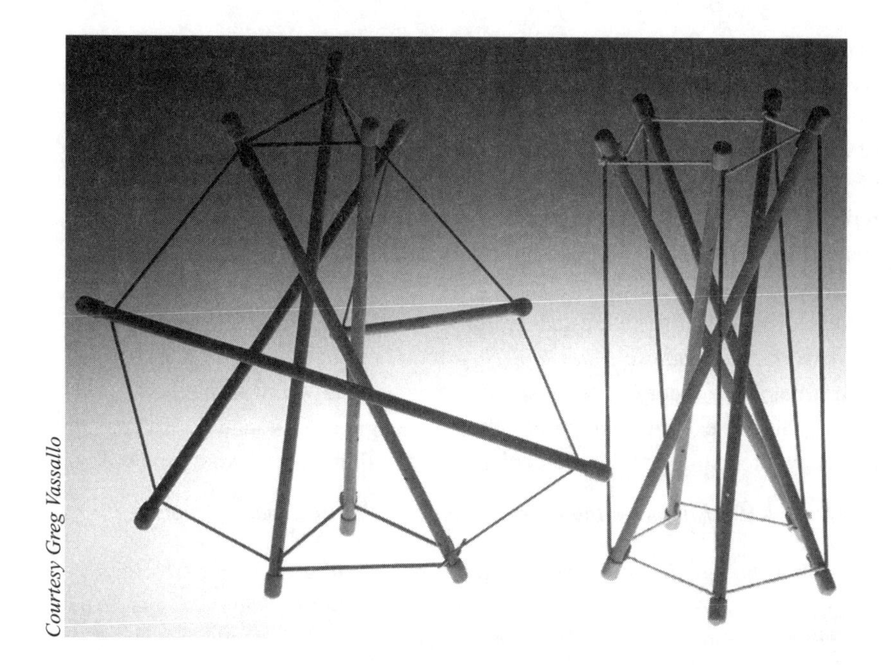
Courtesy Greg Vassallo

FIGURE 4.32

kit for building tensegrity frameworks, "Tensegritoy." Although tensegrity frameworks are easily constructed, the theory of tensegrity frameworks is not so well understood.

We can draw some conclusions similar to those we made for dimension two. Clearly, we can eliminate vertices of valence 1 or 2. But what about vertices of valence 3? Can we eliminate them, just as we eliminated vertices of valence 2 in the planar case? The planar argument against two rods at a point can be extended, permitting us to exclude three rods (and no cables) at a vertex as being unstable, or as being unstressed, and therefore removable. Three cables at a vertex may also be eliminated, as outlined in the next exercise. At first glance, two rods and one cable seem to be unstable. However, two cables and a rod definitely cannot be excluded, as the first framework in Figure 4.32 illustrates. Thus the same lower bound, $|E| \geq \frac{3}{2}|V|$, on the number of rods and cables that we had in the 2-dimensional case holds in 3-space. Furthermore, this bound cannot be improved upon; there are tensegrity frameworks with the average valence as close to 3 as you wish.

One can understand why this is true by taking a closer look at the first framework in Figure 4.32. Visualize removing, one at a time, the rods that connect two 3-valent vertices allowing the cable through these 3-valent vertices to "straighten out." The result will be a much simpler rigid tensegrity framework consisting of three rods and three cables connecting two triangles of cables. This is the triangular prism; the right-hand model is the pentagonal prism. Once we have a prism or any rigid tensegrity framework we may add rods between two cables, increasing the number of 3-valent vertices by two. Repeating this construction enough times will bring the edge to vertex ratio as close to $\frac{3}{2}$ as you wish.

One place for the interested reader to find out more about tensegrity frameworks is *Geodesic Math and How to Use It* by Hugh Kenner. Among other things, the theory of prisms is completely worked out in his book.

This closing exercise, which is rather challenging, represents just a few of the many interesting problems associated with building tensegrity frameworks.

Exercise 4.10. *Eliminate the need for joints with three rods or three cables.*

1. *Show that, if the three rods or three cables, or two rods and a single cable, or two cables and a single rod do not lie in a common plane, then they cannot be under stress. In this case, they are simply appendages to rigid tensegrity frameworks and can be deleted without having any effect on that framework.*

2. *Show that three rods that lie in a common plane are unstable.*

3. *Discuss the conditions under which two rods and a single cable, or two cables and a single rod, that lie in a common plane will be stable or unstable.*

4. *Show that three cables that lie in a common plane can be replaced by a triangle of cables joining the other end joints of the three cables.*

Further Reading:
An Annotated Bibliography

RIGIDITY

No undergraduate level textbook on rigidity theory exists, but several books and popular articles illumine specific aspects of rigidity.

The General Theory

Crapo, Henry. "Structural Topology, or the Fine Art of Rediscovery." *The Mathematical Intelligencer* 19:4 (Fall 1997) 27–34. Provides a history of the journal *Structural Topology* (see below) and the research group that published it.

Dewdney, A. K. "The Theory of Rigidity, or How to Brace Yourself Against Unlikely Accidents" (Mathematical Recreations). *Scientific American* 264:5 (May 1991) 126–8.

Graver, Jack, Brigitte Servatius and Herman Servatius. *Combinatorial Rigidity*. Graduate Studies in Mathematics, vol. 2. Providence, R.I.: American Mathematical Society, 1993. A graduate level textbook, it includes a very extensive annotated bibliography of research papers in the area.

Structural Topology. Published from 1978 through 1997 in 24 issues, this journal is a major source of interesting articles on rigidity and related subjects. If you can find a library that subscribed to it, you will surely see many articles of interest as you leaf through the issues.

Historical Resources

These are the works cited in Section 4.1, "A Short History of Rigidity."

Bricard, R. "Mémoir sur la Théorie de l'Octaédre Articulé." *J. Math. Pure and Appl.* 5:3 (1897) 113–48.

Cauchy, A. L. "Sur la Polygons et les Polyédres." *Second Memoire, J. École Polytechnique* 19 (1813) 87–98.

Euler, L. *Opera Postuma*, vol. 1. Petropoli, 1862, 494–6.

Gluck, H. "Almost All Simply Connected Closed Surfaces Are Rigid." In *Geometric Topology.* L. C. Glaser and T. B. Rushing, eds. Springer Lecture Notes, no. 438. New York: Springer, 1975, 225–39.

Henneberg, L. *Die graphische Statik der starren Systeme.* Leipzig, 1911; Johnson Reprint, 1968.

Laman, G. "On Graphs and Rigid Plane Skeletal Structures." *J. Engrg. Math.* 4 (1970) 331–40.

RELATED TOPICS

Linkages

Kempe, A. B. *How to Draw a Straight Line.* Reprint edition: Classics in Mathematics Education Series. Reston, Va.: National Council of Teachers of Mathematics, 1997. Anyone interested in linkages should read this clear account, based on a lecture first published in 1877.

Thurston, William, and Jeffrey Weeks. "The Mathematics of Three-Dimensional Manifolds." *Scientific American* 251:1 (July 1984) 108–120. An application of linkages to topology is developed.

Geodesic Domes

Many books and articles relating to geodesic domes are available. For a very readable account of the technical details, consult:

Kenner, Hugh. *Geodesic Math and How to Use It.* Berkeley: University of California Press, 1976. This book includes a short introduction to tensegrity but is mainly devoted to a detailed study of the mathematics needed to design geodesic domes.

Fullerenes and Nanotechnology

Collins, Philip, and Phaedon Avouris. "Nanotubes: The Future of Electronics." *Scientific American* 283:6 (December 2000) 62–69.

Fowler, P. W., and D. E. Manolopoulos. *An Atlas of Fullerenes.* International Series of Monographs on Chemistry, no. 30. Oxford: Clarendon, 1995.

Koruga, Djuro, Stuart Hameroff, James Withers et al. *Fullerene C_{60}: History, Physics, Nanobiology, Nanotechnology.* Amsterdam and New York: North-Holland, 1993. The history of fullerenes is well recounted here, and its bibliography covers the chemistry papers mentioned in Chapter 4.

Mirsky, Steve. "Tantalizing Tubes." *Scientific American* 282:6 (June 2000) 40–42.

Tensegrity

Connelly, Robert, and Allen Beck. "Mathematics and Tensegrity." *American Scientist* 86 (March–April, 1998) 142-51.

Ingber, Donald E. "The Architecture of Life." *Scientific American* 278:1 (January 1998) 48–57. Discusses applications of tensegrity to the life sciences.

Pugh, Tony. *An Introduction to Tensegrity.* Berkeley: University of California Press, 1976.

GENERAL BACKGROUND READING

Graph Theory

There are many good introductory books on graph theory, far too many to list here. The following selections are very readable.

Baglivo, Jenny A. and Jack E. Graver. *Incidence and Symmetry in Design and Architecture.* Cambridge University Press, 1983. The first half looks at graph theory, while the second half introduces group theory and symmetry.

Ore, Oystein. *Graphs and Their Uses.* New Mathematical Library Series, vol. 34. Washington, D.C.: Mathematical Association of America, 1990. This old standby is a very inexpensive paperback.

Roberts, Fred. *Discrete Mathematical Models*. Englewood Cliffs, N.J.: Prentice Hall, 1976. An introduction to graphs and directed graphs, this book includes a wide variety of applications.

Tucker, Alan. *Applied Combinatorics*. New York: Wiley, 1995. A fine introduction to graph theory, followed by an exposition of enumeration theory.

Geometry

Very many books address polyhedral geometry. Here are a few that I found useful.

Coxeter, H. S. M. *Regular Polytopes*. New York: Dover, 1973.

Cundy, H. Martyn, and A. P. Rollett. *Mathematical Models*. Oxford University Press, 1961.

Kappraff, Jay. *Connections: The Geometric Bridge Between Art and Science*. New York: McGraw-Hill, 1991.

Loeb, Arthur L. *Space Structures: Their Harmony and Counterpoint*. Reading, Mass.: Addison-Wesley, 1976.

Seneschal, Marjorie, and George Fleck, eds. *Shaping Space*. Boston: Birkhäuser, 1988.

Wenninger, Magnus J. *Polyhedron Models*. Cambridge University Press, 1971.

Software, Model Kits and the Web

Commercial software packages for making geometric constructions are available for both Windows and Macintosh computers. *Geometer's Sketchpad*, a product of Key Curriculum Press, has a wealth of support materials both in print and on the web. *Cabri*, marketed in the United States by Texas Instruments, is also available on their hand-held TI-92 calculator. Many more downloadable specialized construction programs are available on the web.

All of the tensegrity models pictured in this book were made with a construction kit from Design Science Toys called "Tensegritoy." Ordering information and details are available on the web, as well as information about a variety of other construction kits.

Growing numbers of websites are devoted to one or more of the topics in this book. With the sites changing daily, printed lists are quickly outdated. The best way to find current websites is to employ a search engine.

Index